Low Pressure Boilers
fifth edition

Study Guide

atp AMERICAN TECHNICAL PUBLISHERS
Orland Park, Illinois

Frederick M. Steingress

Technical information and assistance was also provided by the following companies:

Bell & Gossett
Cleaver-Brooks
Honeywell Inc.

5 6 7 8 9 – 19 – 9 8 7 6 5 4 3 2

Printed in the United States of America

ISBN 978-0-8269-4373-6

This book is printed on recycled paper.

Contents

Low Pressure Boilers

Introduction

Low Pressure Boilers Study Guide is designed to reinforce information presented in *Low Pressure Boilers*. The textbook is used as a reference to complete the learning activities in the study guide. Each chapter in the study guide covers information in the corresponding chapter in the textbook. Chapter 13 – Boiler Operator Licensing contains seven separate exams that can be used as a comprehensive review. The questions in Chapter 13 are similar to the type of questions found on a typical boiler operator licensing examination.

Answering Chapter Questions

The question types used in the study guide include true-false, multiple choice, matching, and essay. For true-false questions, circle the letter T if the statement is true, or circle the letter F if the statement is false. For multiple choice questions, place the letter of the correct answer in the answer blank next to the question. For matching questions, place the letter of the correct corresponding answer in the answer blank next to the question. For essay questions, write the answers on a separate sheet of paper and submit to the instructor.

Related Information

Information presented in *Low Pressure Boilers* and the *Low Pressure Boilers Study Guide* addresses common boiler operation topics. Additional information related to boiler operation/stationary engineering is available in other American Technical Publishers products. To obtain information about these products, visit the American Technical Publishers website at www.atplearning.com.

The Publisher

Chapter 1

Boiler Operation Principles

Section 1.1—Boilers in Industry

Name ______________________ **Date** ______________

True-False

T F **1.** Steam is the vapor that forms when water is heated to its boiling point.

T F **2.** A boiler is an open metal container in which water is heated to produce steam or heated water.

T F **3.** The greatest impact of the use of steam power occurred during the Industrial Revolution of the eighteenth century and early nineteenth century.

Multiple Choice

______________ **1.** ___ is commonly used for heating or process applications.
- A. Coal
- B. Steam
- C. Gas
- D. Feedwater

______________ **2.** ___ applications use steam to produce heat, force, and/or mechanical action in a production application.
- A. Process
- B. Dry-cleaning
- C. Building space heating
- D. Laundry

______________ **3.** ___ is credited with the development of the reciprocating steam engine.
- A. Richard Trevithick
- B. Hero of Alexandria
- C. Thomas Newcomen
- D. James Watt

Chapter 

Boiler Operation Principles

Section 1.2—Boiler Operation Theory

Name ______________________ **Date** ______________

True-False

T F **1.** Steam is produced as water is heated at its boiling point (212°F at atmospheric pressure).

T F **2.** A backflow preventer is a boiler accessory that prevents the flow of water back to the potable water supply.

T F **3.** Steam from a boiler is directed by the main steam line through the main steam stop valve and enters the steam header.

T F **4.** Steam that has given up its heat and turned back to water is boiler water.

T F **5.** Heat is generated in a boiler by the combustion of a fuel such as gas or fuel oil.

T F **6.** An internal furnace is a furnace in a boiler that is surrounded by a heating surface.

T F **7.** Feedwater is water that is treated for use in a boiler.

T F **8.** Heat always flows from a material having a lower temperature to a material having a higher temperature.

T F **9.** Latent heat is heat identified by a change of state and no temperature change of the substance.

T F **10.** Conduction is heat transfer that occurs when molecules in a material are heated and the heat is passed from molecule to molecule through the material.

T F **11.** Approximately 212 lb of air are required to burn a pound of fuel.

T F **12.** A gas law relates to events that occur as heat is generated by the combustion of a fuel and is used to produce steam or hot water in a boiler.

T F **13.** There are four main causes of natural circulation in a boiler.

T F **14.** Perfect combustion is combustion that occurs when all the fuel is burned with a minimum amount of excess air remaining.

T F **15.** Saturated steam is pure steam at a temperature that corresponds to the boiling point of water at a specific pressure.

Multiple Choice

__________ **1.** ___ is heat transfer that occurs when currents circulate between warm and cool regions of a fluid.
A. Combustion
B. Radiation
C. Conduction
D. Convection

__________ **2.** The ___ is the part of a boiler with water on one side and heat from gases of combustion on the other.
A. furnace volume
B. heating surface
C. fire side
D. water side

__________ **3.** Sensible heat is heat that ___.
A. involves a change of state
B. involves no temperature change of the substance
C. changes ice to water
D. can be measured with a thermometer or sensed by a person

__________ **4.** A ___ is used to transfer heat from one substance to another without allowing the materials to mix.
A. thermometer
B. heat exchanger
C. thermostat
D. regulator

__________ **5.** ___ is the form of energy identified by a temperature difference or a change of state.
A. Gas
B. Steam
C. Heat
D. Vapor

__________ **6.** ___ combustion is combustion that occurs when fuel is burned using only the theoretical amount of air.
A. Perfect
B. Incomplete
C. Complete
D. Primary

______________ **7.** ___ air is air supplied to a burner above the theoretical amount required to burn the fuel to ensure complete combustion.

A. Primary
B. Excess
C. Secondary
D. Pilot

______________ **8.** ___ is the process of reducing fuel oil into a fine spray of minute particles.

A. Steam priming
B. Convection loading
C. Atomization
D. Discharge shaping

______________ **9.** A ___ is used to reduce utility water pressure to prevent overpressure in a system.

A. steam outlet
B. safety valve
C. pressure regulator
D. water level

______________ **10.** During operation, a ___ holds water, transfers heat to the water to make steam, and collects the steam that is produced.

A. draft system
B. steam system
C. feedwater pump
D. steam boiler

______________ **11.** The Fahrenheit scale uses 0°F as the freezing point of salt water and provides a larger number of increments between the freezing point of pure water (32°F) and boiling point of pure water (___°F) compared to the Celsius scale.

A. 100
B. 150
C. 200
D. 212

______________ **12.** The amount of energy required to raise the temperature of 1 lb of water 1°F is 1 ___.

A. NOWL
B. therm
C. psia
D. Btu

______________ **13.** Heat is transferred by ___.

A. conduction, convection, and enthalpy
B. convection, radiation, and enthalpy
C. radiation, steam, and condensate
D. conduction, convection, and radiation

_________________ **14.** ___ is the process that occurs when enough heat is added or removed to change a substance from one physical form to another.
A. Thermodynamics
B. Change of state
C. Atomization
D. Combustion

_________________ **15.** Thermal efficiency can be as high as ___% on some boilers.
A. 65
B. 80
C. 95
D. 100

Steam Heating Systems

_________________ **1.** Condensate receiver tank

_________________ **2.** Steam line

_________________ **3.** Feedwater pump

_________________ **4.** Steam header

_________________ **5.** Branch lines

_________________ **6.** Main steam stop valve

_________________ **7.** Heating unit

_________________ **8.** Steam trap

_________________ **9.** Boiler

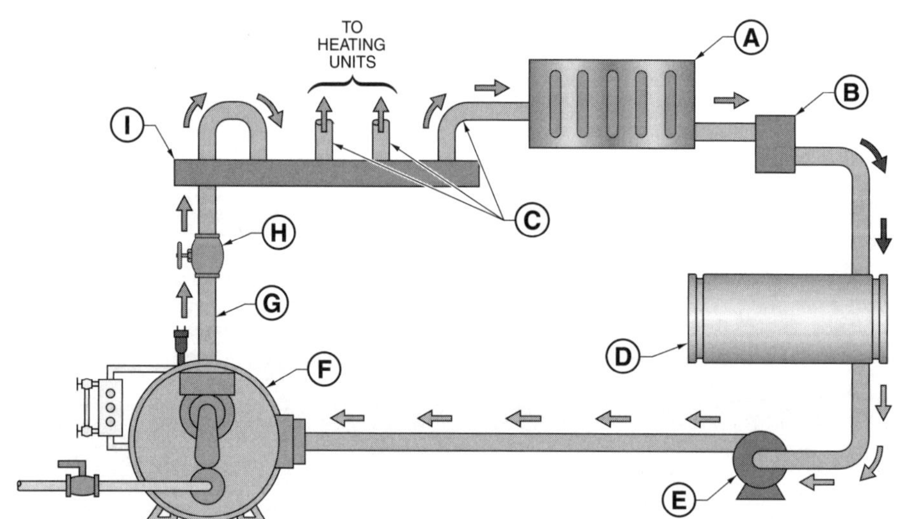

Chapter 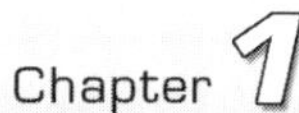

Boiler Operation Principles

Section 1.3—Boiler Systems

Name ______________________________ **Date** ____________________

True-False

T F **1.** For fuel to burn efficiently, the proper amount of oxygen must be provided.

T F **2.** A steam system is a boiler system that supplies the proper amount of water to a boiler.

T F **3.** Hot water boilers have a hot water distribution system instead of a steam system.

T F **4.** Boiler systems are commonly depicted using drawings, diagrams, and symbols.

T F **5.** A feedwater system is a boiler system that collects, controls, and distributes the steam produced in a boiler.

Multiple Choice

______________ **1.** A ___ system provides the air necessary for combustion.
- A. feedwater
- B. draft
- C. steam
- D. fuel

______________ **2.** The four systems necessary to operate a boiler are feedwater, fuel, draft, and ___.
- A. combustion
- B. stoker
- C. condensate
- D. steam

______________ **3.** Boilers used in industry are designed to ___.
- A. minimize the cost of producing steam
- B. generate utility power
- C. heat schools and hospitals
- D. monitor building temperatures

Boiler Systems

______________ **1.** Steam system

______________ **2.** Boiler

______________ **3.** Feedwater system

______________ **4.** Draft system

______________ **5.** Fuel system

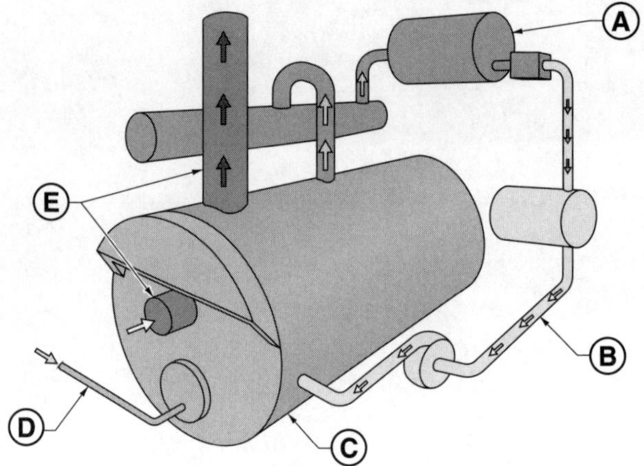

Chapter 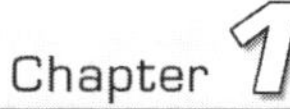

Boiler Operation Principles

Section 1.4—Boiler Design and Construction

Name ______________________________ **Date** ____________________

True-False

T F **1.** In a scotch marine boiler, the gases of combustion pass through tubes that are surrounded by water.

T F **2.** The ASME Code governs boiler design, material, method of construction, inspection, and quality assurance.

T F **3.** A low pressure boiler is a boiler that has an MAWP over 15 psi.

T F **4.** A package boiler is a self-contained unit that is preassembled at a factory.

T F **5.** A firebox boiler is a cylindrical scotch marine boiler.

T F **6.** A tube bank eliminates the need for a seal-welded inner casing.

T F **7.** A high pressure boiler is a boiler with an MAWP of more than 15 psi.

T F **8.** Atmospheric pressure is the sum of gauge pressure and vacuum pressure.

Multiple Choice

______________ **1.** A ___ boiler is a boiler in which heat and gases of combustion pass through tubes surrounded by water.
- A. firetube
- B. watertube
- C. cast iron
- D. straight-tube

______________ **2.** A ___ boiler is a boiler in which water passes through tubes surrounded by gases of combustion.
- A. firetube
- B. watertube
- C. cast iron
- D. firebox

________________ **3.** ___ are used in boilers to direct the gases of combustion for multiple passes to obtain maximum efficiency.

A. Combustion controls
B. Firetubes
C. Baffles
D. Zone controls

________________ **4.** A boiler's ___ is the highest pressure in pounds per square inch (psi) at which it can be safely operated.

A. NOWL
B. ASME
C. MAWP
D. LWFC

________________ **5.** A ___ boiler does not use tubes.

A. cast iron
B. scotch marine
C. watertube
D. firebox

________________ **6.** ___ pressure is the pressure caused by the weight of air surrounding the earth.

A. Atmospheric
B. Thermal
C. Conductive
D. Absolute

________________ **7.** Because each section of a(n) ___ boiler is modular, sections can be assembled on-site to produce the boiler capacity required.

A. firetube
B. industrial watertube
C. scotch marine
D. cast iron

________________ **8.** A(n) ___ is a component directly attached to a boiler that is required for the operation of the boiler.

A. fitting
B. accessory
C. fuel oil tank
D. condensate return tank

________________ **9.** The nonflammable material used to insulate the outer surface of a boiler from heat is ___.

A. refractory
B. a steam bridge
C. an aquastat
D. a tube bank

_______________ **10.** A ___ boiler is a boiler that uses the municipal solid waste that would normally go to landfills.

A. cast iron
B. refuse
C. waste
D. firebox

Scotch Marine Boilers

_______________ **1.** Tubes

_______________ **2.** Tube sheet

_______________ **3.** Gases of combustion

_______________ **4.** Internal furnace

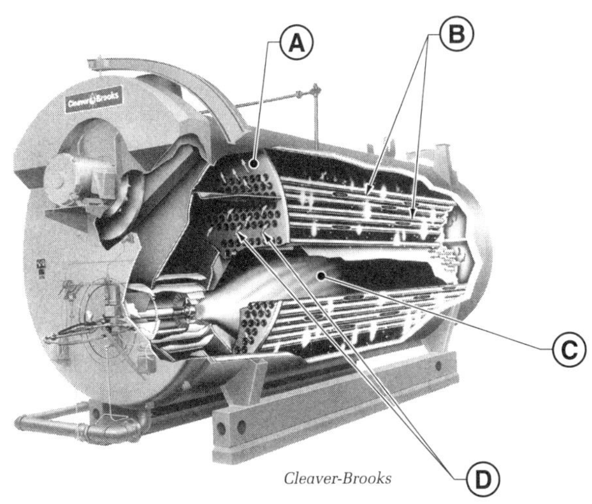

Cleaver-Brooks

Straight-Tube Watertube Boilers

______________ **1.** Steam and water drum

______________ **2.** Internal feedwater line

______________ **3.** Tubes

______________ **4.** Gases of combustion

______________ **5.** Furnace

______________ **6.** Burner assembly

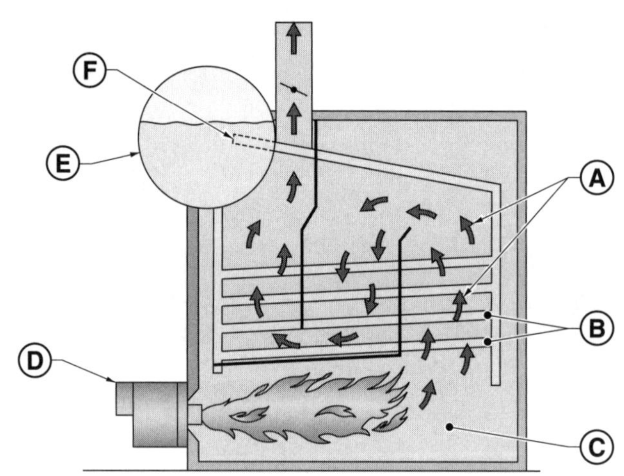

Steam Boiler Fittings

Section 2.1—Safety Valves

Name ______________________ **Date** ______________

True-False

T F **1.** The MAWP for low pressure steam boilers is 15 psi.

T F **2.** Safety valves on steam boilers are designed to open slowly.

T F **3.** During normal operation, safety valves are designed to close quickly without chattering.

T F **4.** Safety valves are tested only by performing a try lever test.

T F **5.** A safety valve try lever test is performed with the boiler at a minimum of 15 psi or higher.

T F **6.** The *ASME Boiler and Pressure Vessel Code* states that a pop test should be performed semiannually.

T F **7.** A safety valve's capacity is listed on the data plate attached to the safety valve.

T F **8.** When the total force applied on a safety valve disc exceeds the spring pressure, the valve disc is lifted off its seat.

T F **9.** In addition to safety valves, four other types of valves are sometimes used on boilers to prevent overpressure.

T F **10.** A faulty safety valve could cause a boiler explosion.

Multiple Choice

______________ **1.** According to the *ASME Boiler and Pressure Vessel Code,* the ___ is the most important valve on a boiler.

A. main steam stop valve
B. safety valve
C. automatic nonreturn valve
D. feedwater stop valve

_______________ **2.** According to the ASME Code, the MAWP of a low pressure steam boiler is ___ psi or less.
A. 10
B. 15
C. 100
D. 212

_______________ **3.** The total force on a safety valve is equal to the ___.
A. area multiplied by diameter
B. area multiplied by distance
C. area multiplied by pressure
D. MAWP multiplied by pressure

_______________ **4.** If a safety valve is 3″ in diameter, its area is ___ sq in.
A. 2.05
B. 3.14
C. 4.32
D. 7.07

_______________ **5.** The *ASME Boiler and Pressure Vessel Code* allows ___ valve(s) to be placed between a boiler and a safety valve or between a safety valve and the point of discharge.
A. no
B. one
C. two
D. three

_______________ **6.** Safety valve capacity is the amount of ___, in pounds per hour (lb/hr), that the safety valve is capable of venting at the rated pressure of the safety valve.
A. water
B. gas
C. steam
D. vapor

_______________ **7.** The total force on a safety valve 3.5″ in diameter with a steam pressure of 14 psi is ___ lb.
A. 110.3
B. 122.0
C. 134.7
D. 145.9

_______________ **8.** A ___ valve must stay open until there is a drop in pressure of about 2 psi to 4 psi below the opening pressure.
A. pop test
B. safety
C. chattering
D. huddling

_______________ **9.** Blowback is the ___ in a boiler that occurs after the safety valve has opened.
A. drop in pressure
B. carryover
C. increase of pressure
D. chattering of a feedwater valve

_______________ **10.** According to the ASME Code, a try lever test on a safety valve should be performed every ___ the boiler is in operation and after any period of inactivity.
A. hour
B. 4 hours
C. 7 days
D. 30 days

_______________ **11.** The purpose of a safety valve is to prevent the pressure in a boiler from ___.
A. exceeding the MAWP
B. dropping below the MAWP
C. causing a furnace explosion
D. relieving water pressure

_______________ **12.** The capacity of a safety valve is measured by the amount of steam that can be discharged per ___.
A. shift
B. hour
C. minute
D. day

_______________ **13.** ___ is the rapid opening and closing of a safety valve.
A. Depressurizing
B. Pressurizing
C. Chattering
D. Huddling

_______________ **14.** The safety valve on a low pressure boiler is designed to open when pressure in the boiler exceeds the ___.
A. safety valve setting
B. NOWL
C. feedwater pump pressure
D. bottom blowdown discharge pressure

_______________ **15.** After the total force of steam has lifted a safety valve off its seat, the steam enters the ___.
A. huddling chamber
B. combustion chamber
C. steam holding tank
D. main steam line

Safety Valves

______________ **1.** Valve disc

______________ **2.** Body

______________ **3.** Try lever pin

______________ **4.** Try lever

______________ **5.** Spindle

______________ **6.** Spring

______________ **7.** Valve seat

______________ **8.** Bonnet

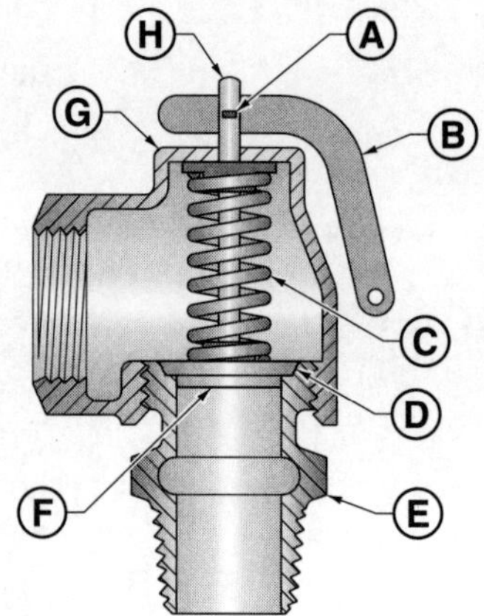

Safety Valve Huddling Chambers

______________ **1.** Valve seat

______________ **2.** Huddling chamber

______________ **3.** Spring

______________ **4.** Valve spindle

______________ **5.** Steam pressure

______________ **6.** Valve disc

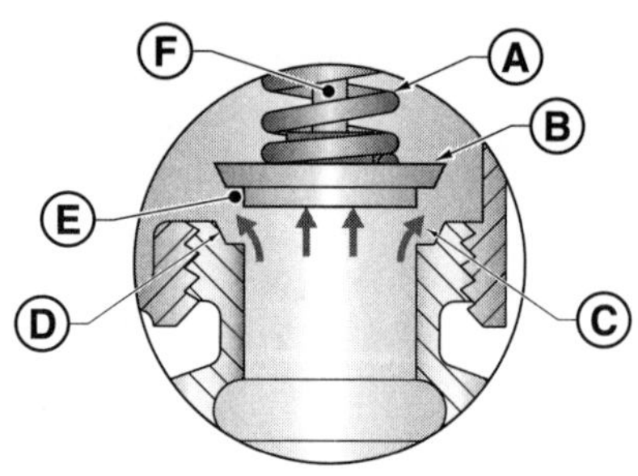

Steam Boiler Fittings

Section 2.2—Steam Pressure Gauges

Name ______________________________ Date ____________________

True-False

T F **1.** A steam pressure gauge must be connected to the highest part of the steam side of a boiler and must be positioned to allow easy viewing by the boiler operator.

T F **2.** A siphon is connected between a boiler and a steam pressure gauge to prevent water from entering the Bourdon tube.

T F **3.** A slow gauge indicates less pressure than is actually present in the boiler.

T F **4.** Steam allowed to enter the Bourdon tube of a pressure gauge can result in a false reading and/or cause damage from warping.

T F **5.** The effect of hydrostatic pressure must be considered when calibrating a pressure gauge.

T F **6.** The closed end of a Bourdon tube is attached to the steam side of a boiler.

Multiple Choice

______________ **1.** Two common types of steam siphons are the ____ siphons.
- A. Bourdon tube and pigtail
- B. pigtail and U-tube
- C. spiral and S-tube
- D. spiral and U-tube

______________ **2.** A steam pressure gauge on a boiler is calibrated to read ___.
- A. inches of vacuum
- B. pounds per square inch
- C. absolute pressure
- D. pressure below atmospheric pressure

______________ **3.** A(n) ___ prevents steam from entering the Bourdon tube of a pressure gauge.
- A. automatic nonreturn valve
- B. inspector's test cock
- C. os&y valve
- D. siphon

______________ **4.** A ___ pressure gauge can read either vacuum or pressure.
A. compound
B. duplex
C. suction
D. vacuum

______________ **5.** Vacuum is pressure ___ pressure.
A. above gauge
B. below absolute
C. below atmospheric
D. equal to gauge

______________ **6.** The most common steam pressure gauge operates by the movement of a ___.
A. Bourdon tube
B. water column
C. siphon
D. pressure control

______________ **7.** A(n) ___ gauge is a pressure gauge that indicates more pressure than is actually present.
A. steam pressure
B. slow
C. uncalibrated
D. fast

______________ **8.** A ___ is a device that sends a signal to a controller, such as a programmable logic controller or other digital controller.
A. translator
B. relay
C. transmitter
D. dispatch

Steam Pressure Gauge Operation

______________ **1.** Linkage

______________ **2.** Siphon connection

______________ **3.** Steam pressure

______________ **4.** Face

______________ **5.** Needle

______________ **6.** Bourdon tube

______________ **7.** Case

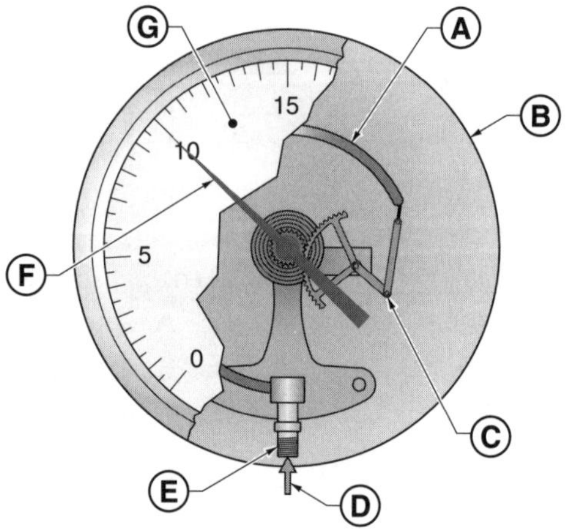

Chapter 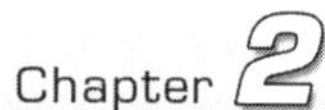

Steam Boiler Fittings

Section 2.3—Water Columns

Name ______________________________ **Date** ____________________

True-False

T F **1.** The water column of a boiler should be blown down each shift.

T F **2.** A water column is used to indicate the water level in a blowdown tank.

T F **3.** Blowdown of a gauge glass is performed by opening the gauge glass blowdown valve.

T F **4.** The purpose of the water column of a boiler is to reduce water turbulence, allowing the true boiler water level to be indicated by the water level in the gauge glass.

T F **5.** In low pressure steam boilers, the lowest visible part of the gauge glass must be placed a minimum of 1″ above the lowest permissible water level.

T F **6.** When blowing down the water column of a boiler, the water column blowdown valve is opened for 5 sec to 10 sec, allowing water and any sludge and/or sediment to be discharged.

T F **7.** If the top line to a gauge glass is closed or clogged, the glass will fill with water.

T F **8.** If the bottom line to a gauge glass is closed or clogged, the glass will be empty.

T F **9.** The *ASME Boiler and Pressure Vessel Code* requires at least one water column on all boilers.

T F **10.** The condition of water column piping is determined by removing boiler cross tee plugs.

Multiple Choice

______________ **1.** The level of the water in a ___ indicates the water level in the boiler.
A. condensate return tank
B. feedwater tank
C. gauge glass
D. blowdown tank

_______________ **2.** A low pressure steam boiler must have ___ or more gauge glasses to determine the water level in the boiler.
A. one
B. two
C. three
D. four

_______________ **3.** A ___ is a pipe fitting with removable plugs that allows for cleaning and inspection of piping.
A. baffle
B. boiler cross tee
C. vent
D. diverter fitting

_______________ **4.** ___ is the process of evacuating water and undesirable accumulated material from a boiler or fitting, such as a gauge glass.
A. Foaming
B. Priming
C. Carryover
D. Blowdown

Water Columns and Gauge Glasses

_______________ **1.** Water column

_______________ **2.** Water column blowdown valve

_______________ **3.** Cross tees

_______________ **4.** Vent

_______________ **5.** Try cocks

_______________ **6.** Isolation valve

_______________ **7.** Gauge glass

_______________ **8.** NOWL

_______________ **9.** Gauge glass blowdown valve

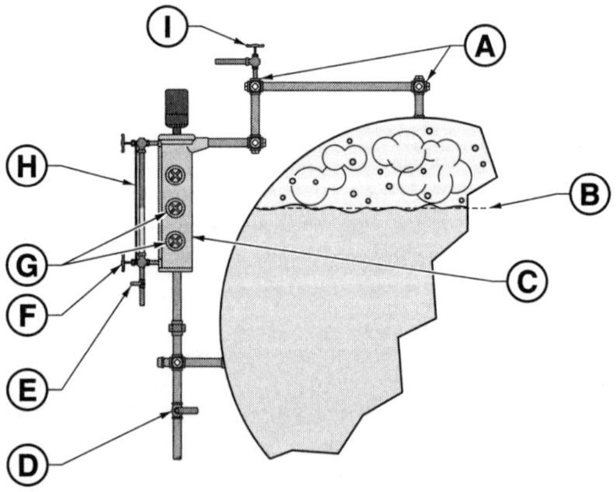

Chapter 

Steam Boiler Fittings

Section 2.4—Blowdown Valves

Name ______________________________ **Date** ____________________

True-False

T F **1.** A bottom blowdown line is piping located at the highest part of the steam side of a boiler.

T F **2.** Increased water surface tension on boiler water can lead to foaming, causing rapid fluctuations in water level.

T F **3.** Boiler water never discharges at the boiler temperature and pressure.

T F **4.** Boilers equipped with a quick-opening valve and a slow-opening valve must have the quick-opening valve closest to the boiler shell.

T F **5.** Only watertube boilers have bottom blowdown valves.

T F **6.** A high water level in a boiler can lead to carryover.

Multiple Choice

______________ **1.** A ___ is a vessel that prevents damage to a sewer system from steam and hot water by holding blowdown water until it is cool enough for safe discharge.

A. pressure control
B. feedwater tank
C. gauge glass
D. blowdown tank

______________ **2.** Surface impurities increase the surface tension in a blowdown line by creating a film which prevents ___ bubbles from breaking through the surface of the water.

A. air
B. oxygen
C. CO_2
D. steam

______________ **3.** The best time to blow down a boiler to remove sludge and sediment is when the boiler is under a ___.

A. high load
B. light load
C. load twice the MAWP
D. maximum load

__________________ **4.** When performing a bottom blowdown, the valve closest to the boiler is always opened ___.
A. first and closed first
B. first and closed last
C. last and closed first
D. last and closed last

__________________ **5.** ___ added to boiler water are used to turn scale-forming salts into a nonadhering sludge that stays in suspension in the boiler water.
A. Oxygen
B. Minerals
C. Slag
D. Chemicals

__________________ **6.** ___ are located on the blowdown line used to release water from the bottom of a boiler.
A. Bourdon tubes
B. Bottom blowdown valves
C. Operating controls
D. Pressure controls

__________________ **7.** ___ is a condition that occurs when large slugs of boiler water are carried into steam lines.
A. Priming
B. Carryover
C. Chattering
D. Blowback

__________________ **8.** A(n) ___ is a small tank for boiler blowdown water to flow through, where city water is added to the boiler water to reduce the discharge water temperature.
A. huddling chamber
B. exchanger
C. blowdown separator
D. feedwater tank

Bottom Blowdown Valves

__________________ **1.** Quick-opening valve

__________________ **2.** Slow-opening valve

__________________ **3.** To blowdown line

__________________ **4.** From boiler

Chapter 2

Steam Boiler Fittings

Section 2.5—Boiler Vents

Name ______________________ **Date** ______________

True-False

T F **1.** A boiler vent is located at the highest part of the steam side of a boiler.

T F **2.** All boilers are equipped with a boiler vent.

T F **3.** The boiler vent should be open when filling a boiler with water.

T F **4.** A handhole is a hole in a boiler shell that provides access for a person during maintenance and repair operations.

T F **5.** Cutting a boiler in on-line is the opening of the main steam stop valve to the main steam header to allow steam to flow to its point of use.

Multiple Choice

______________ **1.** To prevent air pressure from building up in a boiler when filling it with water, the ___ must be open.

A. safety valve
B. main steam stop valve
C. boiler vent
D. manhole cover

______________ **2.** A ___ is used to prevent a dangerous vacuum from forming in a boiler when it is taken out of service.

A. safety valve
B. boiler vent
C. main steam stop valve
D. manhole cover

______________ **3.** A ___ provides access inside the water side of a boiler for inspection, sight, or cleaning.

A. boiler vent
B. safety valve
C. handhole
D. siphon

_______________ **4.** At sea level, a vacuum in a boiler results in a force of ___ lb applied on every square inch of surface.
A. 14.7
B. 17.5
C. 19.3
D. 21.4

_______________ **5.** A manhole cover can weigh more than 50 lb, and significant injury and/or damage can occur if it is rapidly pulled in by ___ inside the boiler during removal.
A. blowdown
B. venting
C. chattering
D. a vacuum

_______________ **6.** ___ weakens the metal and shortens the life of a boiler.
A. Pitting
B. Chattering
C. Dumping
D. Venting

_______________ **7.** A ___ is a boiler fitting used to vent air from inside the boiler.
A. vacuum
B. boiler vent (air cock)
C. safety valve
D. feedwater tank

_______________ **8.** When a boiler that has been producing steam is taken off-line, steam in the boiler cools and condenses, creating a ___.
A. pressure drop
B. manhole
C. vacuum
D. blowdown

Chapter 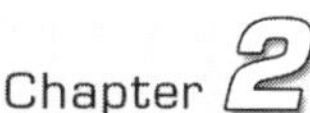

Steam Boiler Fittings

Section 2.6—Pressure Controls

Name ____ **Date** ________________________

True-False

T F **1.** A pressure control is located at the highest part of the steam side of a boiler.

T F **2.** A pressure control must be mounted in a vertical position to ensure accurate operation.

T F **3.** A modulating pressure control regulates the burner firing rate of high fire to low fire by sending an electrical signal that controls a modulating motor connected to a combustion equipment control linkage.

T F **4.** A burner should always be firing for periods shorter than periods when it is off to maintain a consistent furnace temperature and improve efficiency.

T F **5.** An operating range for a boiler is obtained by adjusting the differential pressure setting and either the cut-in pressure setting or the cut-out pressure setting on the pressure control.

T F **6.** A mercury-tube pressure control is a pressure control that regulates a burner control circuit by the flow of mercury in a mercury tube.

T F **7.** The two types of pressure control are additive pressure control and subtractive pressure control.

T F **8.** A typical standard burner operating control package consists of only two separate pressure controls—the operating control and the high limit control.

T F **9.** A low limit control is a pressure control that senses steam pressure and automatically shuts down a burner if the operating control fails.

T F **10.** Not all boilers use modulating control.

Multiple Choice

________________ **1.** For a burner to cut in at 2 psi and cut out at 6 psi, the cut-out pressure should be set at 6 psi and the differential pressure set at ___ psi.

A. 2
B. 4
C. 6
D. 8

_______________ **2.** The operating range of a boiler is controlled by a(n) ___.
A. aquastat
B. vaporstat
C. pressure control
D. pressurestat

_______________ **3.** A(n) ___ regulates the firing rate between high and low fire of a burner.
A. aquastat
B. vaporstat
C. pressure control
D. modulating pressure control

_______________ **4.** With an additive pressure control, cut-in pressure plus differential pressure equals ___ pressure.
A. pulsating
B. cut-out
C. modulating
D. sequential

_______________ **5.** A ___ should always start up in low fire and shut down in low fire.
A. burner
B. pressure control
C. gauge
D. vent

_______________ **6.** A ___ is a switch that starts a burner, controls a firing rate, or shuts down a burner based on steam pressure.
A. safety valve
B. pressure control
C. gauge
D. transmitter

_______________ **7.** The *ASME Boiler and Pressure Vessel Code* requires each automatically fired steam boiler to be protected from overpressure by ___ pressure-actuated control(s).
A. one
B. two
C. three
D. four

_______________ **8.** The mercury tube in a pressure control must be inspected at least ___ for mercury vaporization.
A. biweekly
B. monthly
C. once a year
D. twice a year

Chapter 

Steam Boiler Fittings

Section 2.7—Heat Exchangers

Name ______________________________ Date ________________

True-False

T F **1.** There are many designs of heat exchangers, but the common element is a relatively thin barrier separating the cooling or heating fluid from the process fluid.

T F **2.** There are different heat transfer coefficient values for every different combination of process heating and cooling fluids and for different designs of heat exchangers.

T F **3.** Plate-and-shell exchangers usually consist of a cylindrical shell containing a heating or cooling fluid.

T F **4.** Many heating services use condensate as a heating fluid.

T F **5.** In almost all heating applications, the control valve of a heat exchanger should have a fail-closed action so that the process does not overheat if the valve fails.

T F **6.** The condensate must be removed from a heat exchanger to allow more steam to enter.

T F **7.** A steam trap is often included in a heat exchanger as a way to ensure that the condensate is clean and can be reused.

T F **8.** Usual protection devices used in a heat exchanger to prevent overpressure from thermal expansion are either a thermal expansion safety relief valve or a rupture disc.

Multiple Choice

______________ **1.** Steam at a pressure of 0.0 psig (14.7 psia) has a temperature of ___°F.

A. 100
B. 212
C. 248
D. 298

______________ **2.** A heat exchanger is used to transfer heat from a ___.

A. higher pressure fluid to a lower pressure fluid
B. lower pressure fluid to a higher pressure fluid
C. hotter fluid to a colder fluid
D. colder fluid to a hotter fluid

_______________ **3.** A ___ is an auxiliary device commonly used with boilers to transfer heat from a hotter fluid to a cooler fluid.

A. heat exchanger
B. blowdown tank
C. blowdown attemperator
D. steam trap

_______________ **4.** A(n) ___ is a common type of heat exchanger used to heat the air in a building space.

A. radiator
B. furnace
C. air duct
D. boiler

_______________ **5.** One of the most common types of heat exchangers in industrial applications is a ___ heat exchanger.

A. waste energy
B. pillow plate
C. shell-and-tube
D. plate-and-shell

Chapter 

Steam Boiler Feedwater Systems

Section 3.1—Feedwater Accessories

Name ______________________________ **Date** ________________

True-False

T F **1.** A high water level condition in a boiler could lead to water hammer and line rupture.

T F **2.** A low water level condition in a boiler can result in damage to boiler heating surfaces.

T F **3.** Water in a boiler is heated and turns to condensate.

T F **4.** A check valve allows water to flow in one direction only.

T F **5.** In a feedwater system, the feedwater tank is a large condensate receiver used to collect condensate before it is directed back to the boiler or deaerator.

T F **6.** In a feedwater system, the condensate return tank collects condensate returned from heating units for use in the boiler.

T F **7.** Gate valves are commonly used as feedwater stop valves.

T F **8.** The stop valve on the feedwater line should be located as close to the boiler as practical for isolating the boiler from feedwater accessories during service.

T F **9.** An advantage of using a gate valve for isolation is that it offers little resistance to the flow of fluid through it when it is fully open.

T F **10.** A check valve is used when fluid flow needs to be restricted or regulated through the valve.

T F **11.** The disc and seat of a globe valve can be replaced without removing the valve body from its installed position.

T F **12.** A surge tank generally includes an automatic water makeup system with a float chamber that is located at the minimum water level.

Multiple Choice

______ **1.** The water in a boiler is heated, turns to steam, and leaves the boiler through the ___.
- A. feedwater line
- B. main header
- C. boiler outlet
- D. main branch line

______ **2.** Feedwater is treated and regulated automatically to meet the demand for ___.
- A. condensate
- B. steam
- C. makeup water
- D. cavitation

______ **3.** A(n) ___ valve is normally opened by the operator to isolate the feedwater to the boiler.
- A. condensate return
- B. check
- C. nonreturn
- D. stop

______ **4.** A feedwater ___ valve should be located as close to the shell of the boiler as practical.
- A. check
- B. stop
- C. nonreturn
- D. regulating

______ **5.** Condensate is separated from steam by a ___ that allows condensate, but not steam, to pass into condensate return lines.
- A. check valve
- B. steam trap
- C. water trap
- D. nonreturn valve

______ **6.** A feedwater ___ valve is located next to a feedwater stop valve and prevents feedwater from flowing from the boiler back to the feedwater pump.
- A. return
- B. check
- C. stop
- D. equalizing

______ **7.** A ___ valve is a valve that allows fluid to flow by raising or lowering a valve disc against a seat.
- A. globe
- B. check
- C. gate
- D. bypass

_______________ **8.** In a feedwater system, a ___ supplies the extra capacity to handle changing loads and peak flows of condensate.

A. feedwater regulator
B. pump
C. feedwater tank
D. surge tank

Feedwater Systems

_______________ **1.** Boiler

_______________ **2.** Check valve

_______________ **3.** Condensate return line

_______________ **4.** Condensate return tank

_______________ **5.** Feedwater line

_______________ **6.** Feedwater pump

_______________ **7.** Heating unit

_______________ **8.** Main steam header

_______________ **9.** Main steam line

_______________ **10.** Main steam stop valve

_______________ **11.** Riser

_______________ **12.** Stop valve

_______________ **13.** Surge tank

_______________ **14.** Steam trap

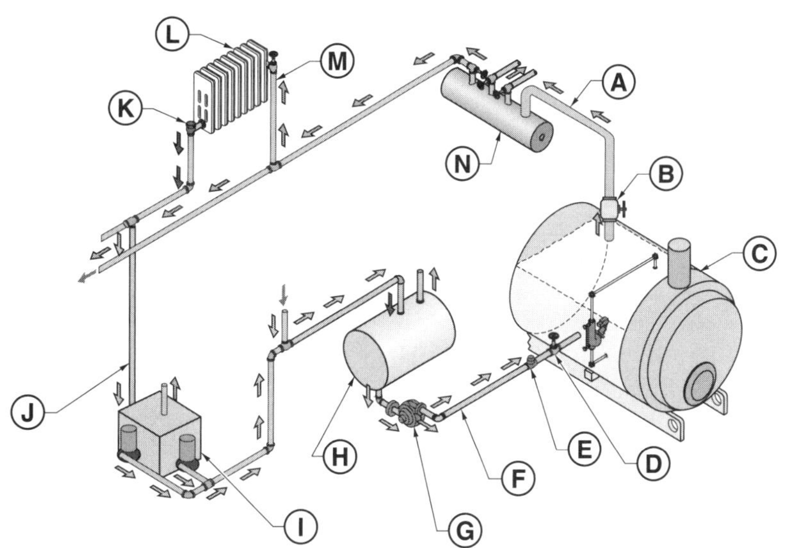

Feedwater Lines

______________________ **1.** Check valve

______________________ **2.** Main feedwater line

______________________ **3.** Boiler

______________________ **4.** From feedwater heater

______________________ **5.** Stop valve

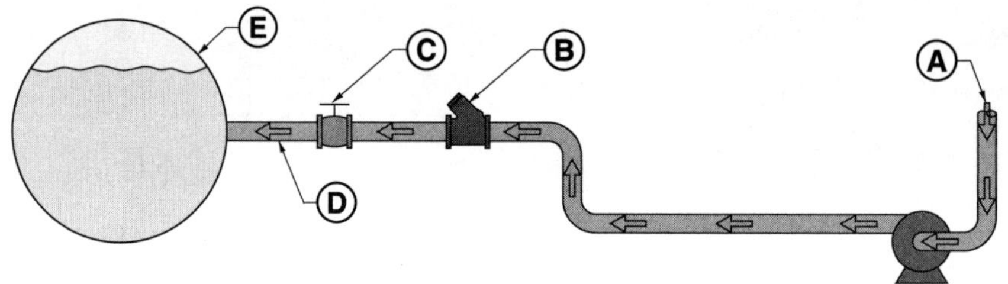

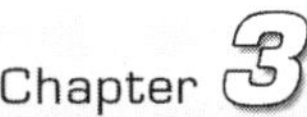

Steam Boiler Feedwater Systems

Section 3.2—Makeup Water Systems

Name ______________________________ **Date** ____________________

True-False

T F **1.** Makeup water is always added from a city water supply line with a manual makeup water system.

T F **2.** A makeup water feeder can be used as a feedwater regulator.

T F **3.** Makeup water is untreated and contains some scale-forming salts that can affect boiler heating surfaces.

T F **4.** The function of an automatic makeup water feeder is to replace water that has been lost.

T F **5.** An automatic makeup water system on a small low pressure boiler connected to the city's water supply uses an automatic makeup feeder located slightly below the NOWL.

Multiple Choice

______________ **1.** Water added to a boiler to replace water lost due to leaks or lost condensate is known as ___ water.

A. extra
B. makeup
C. boiler
D. feed

______________ **2.** Cold makeup water reduces overall boiler efficiency because energy is required to ___ the water.

A. vent
B. heat
C. filter
D. reticulate

______________ **3.** An automatic makeup water feeder is equipped with a ___.

A. discharge valve
B. condensate discharge outlet
C. vacuum pump
D. blowdown valve

Makeup Water Systems

______________________ **1.** Automatic makeup water feeder

______________________ **2.** Backflow preventer

______________________ **3.** Blowdown valves

______________________ **4.** Burner

______________________ **5.** Combination low water fuel cutoff/water column

______________________ **6.** Condensate return line

______________________ **7.** Makeup water supply

______________________ **8.** Manual makeup water valve

______________________ **9.** NOWL

______________________ **10.** Strainer

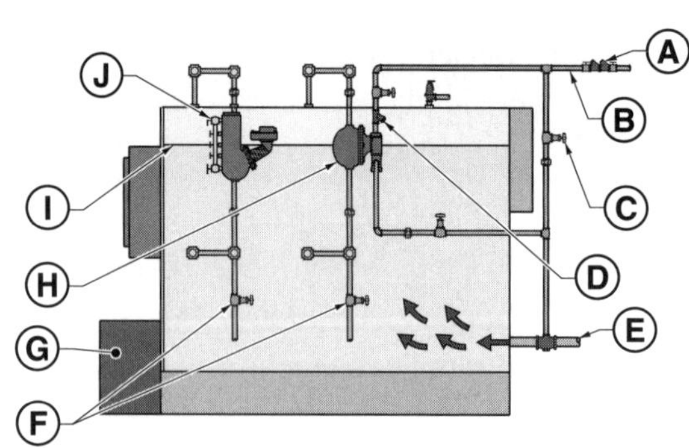

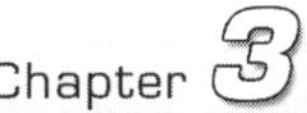

Steam Boiler Feedwater Systems

Section 3.3—Feedwater Regulators

Name ______________________________ **Date** ____________________

True-False

T F **1.** In a system with a feedwater regulator, condensate returns through the condensate return line to the condensate return tank.

T F **2.** The feedwater pressure gauge on a small low pressure steam boiler turns the feedwater pump ON or OFF as required to maintain the NOWL.

T F **3.** Although boiler water level is maintained with a feedwater pressure gauge, the boiler water level must still be routinely checked by a boiler operator.

T F **4.** When the water level in a boiler drops, the feedwater regulator starts the feedwater pump.

T F **5.** Without condensate in a surge tank, the low water alarm on the tank should sound.

Multiple Choice

______________ **1.** The feedwater regulator has a sensing element located at the ___.
- A. boiler vent
- B. MAWP
- C. NOWL
- D. bottom blowdown line

______________ **2.** A ___ maintains the NOWL in a boiler by controlling the amount of feedwater pumped to the boiler from the surge tank.
- A. gauge glass
- B. water column
- C. blowdown tank
- D. feedwater regulator

______________ **3.** If the automatic makeup water feeder on the surge tank fails to supply water to the tank, the ___ will shut off the feedwater pump in order to protect the pump.
- A. low water cutoff
- B. safety valve
- C. drain valve
- D. feedwater regulator

Feedwater Control

_________________ **1.** Boiler

_________________ **2.** Check valve

_________________ **3.** Condensate return line

_________________ **4.** Condensate return tank

_________________ **5.** Feedwater pump

_________________ **6.** NOWL

_________________ **7.** Stop valve

_________________ **8.** Surge tank

_________________ **9.** Vent

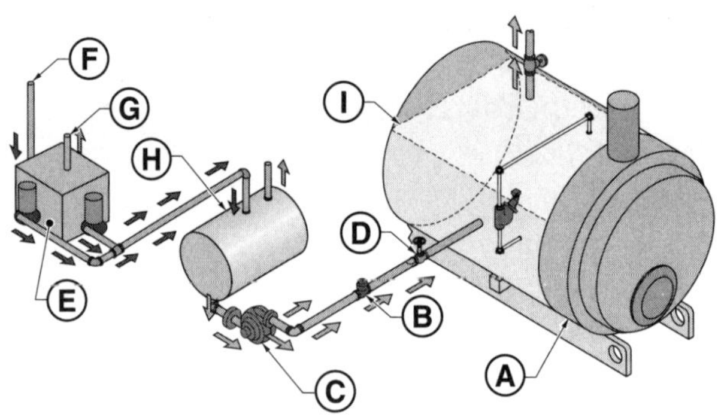

Chapter 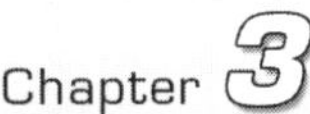

Steam Boiler Feedwater Systems

Section 3.4—Low Water Fuel Cutoffs

Name ______________________________ Date ____________________

True-False

T F **1.** A low water fuel cutoff is located slightly below the NOWL.

T F **2.** An evaporation test is performed by opening a low water fuel cutoff blowdown valve by hand.

T F **3.** A boiler operator must be present during an evaporation test and continually monitor the boiler's water level.

T F **4.** At least once a month, a low water fuel cutoff should be tested with an evaporation test.

T F **5.** The location of an auxiliary low water fuel cutoff allows it to act as a backup if the primary low water fuel cutoff fails to shut the burner OFF in a low water condition.

Multiple Choice

______________ **1.** A probe-type low water fuel cutoff uses ___ to complete a burner control circuit.
A. a float
B. a mercury switch
C. the conductivity of water
D. the water temperature

______________ **2.** The primary function of a low water fuel cutoff is to ___.
A. remove sludge and sediment from the bottom of a boiler
B. shut down a burner if the water level drops below the safe operating level
C. stop the flow of city water supplied to a condensate return tank
D. reduce water turbulence in a gauge glass

______________ **3.** An auxiliary low water fuel cutoff is installed ___.
A. slightly above the NOWL
B. slightly below the primary low water fuel cutoff
C. at the highest point on a boiler
D. on a feedwater line close to the boiler shell

______________ **4.** A(n) ___ shuts the burner OFF in the event of a low water condition.
A. low water alarm
B. feedwater regulator
C. low water fuel cutoff
D. automatic low water feeder

_______________ **5.** A low water fuel cutoff should be tested ___ depending on plant procedures and requirements.
A. daily
B. monthly
C. semiannually
D. annually

Low Water Fuel Cutoff Testing

_______________ **1.** Burner

_______________ **2.** Float at NOWL position

_______________ **3.** Mercury switch

_______________ **4.** Burner control

_______________ **5.** Float chamber

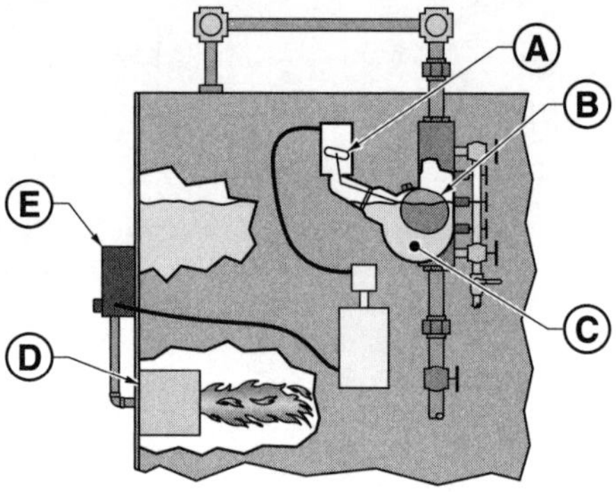

Low Water Fuel Cutoff Parts

_______________ **1.** Float

_______________ **2.** Water column

_______________ **3.** Control switches

_______________ **4.** Steam connection

_______________ **5.** Water connection

_______________ **6.** Float chamber

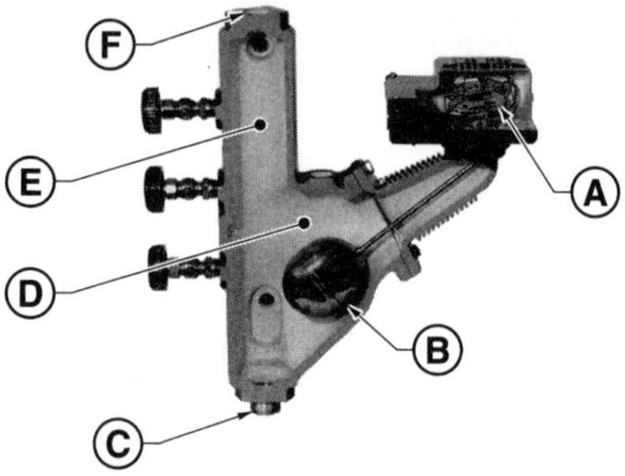

Low Water Fuel Cutoff Operation

_______________ **1.** Feedwater pump turns ON

_______________ **2.** Gauge glass

_______________ **3.** Burner shuts OFF

_______________ **4.** Gauge glass blowdown valve

_______________ **5.** Feedwater pump turns OFF

_______________ **6.** Lowest visible point in gauge glass

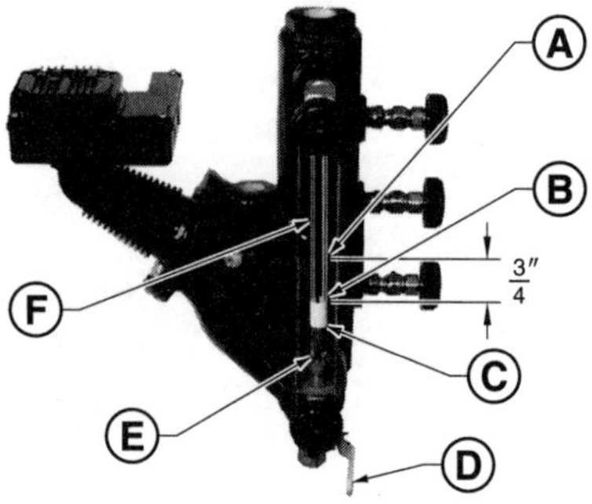

Steam Boiler Feedwater Systems

Section 3.5—Pumps

Name ______________________________ **Date** ____________________

True-False

T F **1.** Suction lift is the pressure at a pump inlet when the liquid supply to the pump is taken from a tank or other supply source that is above the pump.

T F **2.** A reciprocating pump is a pump that operates by adding momentum to a liquid while taking in and discharging fluids in a continuous flow across an unsealed chamber.

T F **3.** The range of pressure on a vacuum switch on a vacuum pump is usually 2″ Hg to 8″ Hg.

T F **4.** A common application of a centrifugal pump in a boiler system is as a feedwater pump.

T F **5.** Gear pumps are widely used positive-displacement pumps because of their simple design and ease of repair.

T F **6.** A vacuum pump is equipped with check valves on both sides of the piston so that the pump moves fluid when the reciprocating component is moving in either direction.

Multiple Choice

__________ **1.** A ___ pump produces a negative pressure on condensate return lines to draw condensate back from a system.

A. fuel oil
B. return
C. gear
D. vacuum

__________ **2.** A typical vacuum pump serving a condensate system is a ___-compartment condensate vessel, divided horizontally.

A. two
B. three
C. four
D. five

______________ **3.** The range of pressure on a vacuum switch is usually ___.
A. 2 psi to 6 psi
B. 6 psi to 12 psi
C. 2″ Hg to 8″ Hg
D. 8″ Hg to 12″ Hg

______________ **4.** The most common type of dynamic pump is a ___ pump.
A. vacuum
B. reciprocating
C. centrifugal
D. positive-displacement

______________ **5.** ___ pumps are often used as fuel-oil pumps and as pumps to add water treatment chemicals to a boiler.
A. Centrifugal
B. Reciprocating
C. Feedwater
D. Positive-displacement

Vacuum Pumps

______________ **1.** Condensate to boiler

______________ **2.** Vacuum tank

______________ **3.** Condensate from system

______________ **4.** Controls

______________ **5.** Vacuum switch

______________ **6.** Pump

______________ **7.** Float-controlled switches

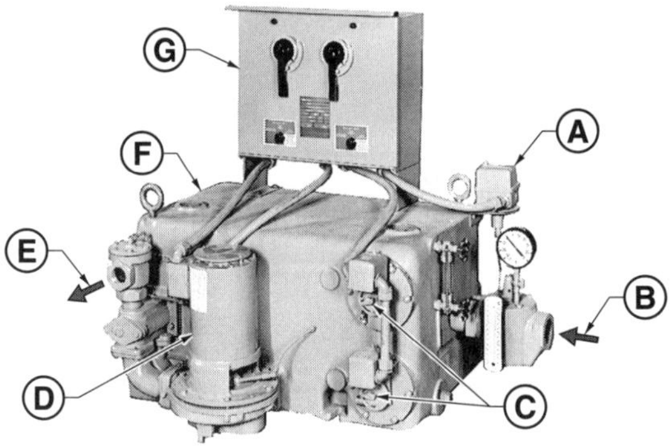

G
F
A
E
B
D
C

Centrifugal Water Pumps

______________ **1.** Feedwater to boiler

______________ **2.** Housing

______________ **3.** Inlet

______________ **4.** Motor

______________ **5.** Outlet

______________ **6.** Rotating impeller

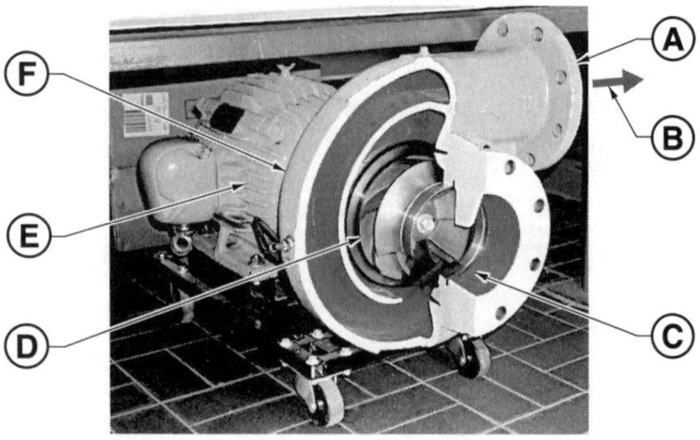

A
F
B
E
C
D

Chapter 

Steam Systems

Section 4.1—Steam Distribution

Name ______________________________ **Date** ____________________

True-False

T F **1.** A nonreturn valve is a type of globe valve that allows a boiler to be cut in on-line automatically when the boiler pressure is at or above the header pressure and allows the boiler to be taken off-line automatically if the pressure in the boiler drops below the header pressure.

T F **2.** Heating units are designed to minimize heat loss from a building space.

T F **3.** A globe valve should never be used as a main steam stop valve.

T F **4.** Gate valves offer no restriction to flow when wide open.

T F **5.** A steam header is a distribution pipe that supplies steam to branch lines.

T F **6.** Expansion bends are used to route steam lines around obstacles.

T F **7.** Steam from a boiler is directed by a main steam line and controlled by a steam header.

T F **8.** An automatic nonreturn valve is constructed of a piston in a cylinder with a stem and handwheel.

T F **9.** Steam headers and branch lines must be supported to prevent damage from strain.

T F **10.** On large boilers, the main steam stop valve can be high above the floor.

Multiple Choice

______________ **1.** A(n) ___ valve is the valve used to cut a boiler in on-line or take a boiler off-line.
- A. automatic nonreturn
- B. main steam stop
- C. os&y gate
- D. safety

______________ **2.** An outside-stem-and-yoke valve is a ___ valve that indicates whether it is open or closed by the position of the stem.
- A. check
- B. globe
- C. gate
- D. safety

__________ **3.** All boilers in battery must have two main steam stop valves or one main steam stop valve and one ___.
- A. steam trap
- B. expansion bend
- C. check valve
- D. automatic nonreturn valve

__________ **4.** When a handwheel is turned to open a nonreturn valve, the stem retracts but the valve ___.
- A. does not automatically open
- B. automatically opens
- C. prevents the flow of steam
- D. drains all condensate

__________ **5.** Even in full open position, a ___ valve causes steam to change direction through the valve, resulting in a drop in steam pressure.
- A. gate
- B. check
- C. safety
- D. globe

__________ **6.** A ___ valve is a type of globe valve that allows a boiler to be cut in on-line automatically when the boiler pressure is at or above the header pressure.
- A. return
- B. nonreturn
- C. test
- D. siphon

Main Steam Stop Valves

__________ **1.** Main steam line

__________ **2.** Condensate

__________ **3.** Water level

__________ **4.** Steam trap

__________ **5.** Main steam stop valve

__________ **6.** Heating unit

__________ **7.** Steam

__________ **8.** Riser

__________ **9.** Main steam header

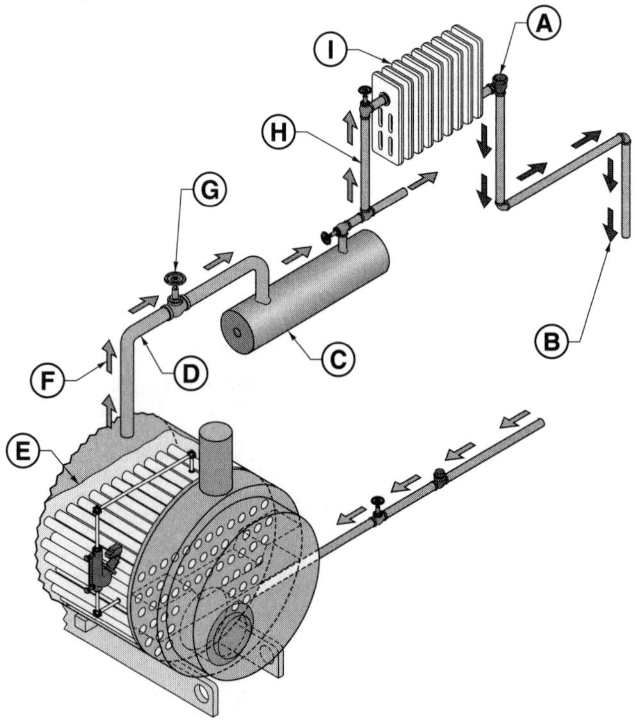

Low Pressure Boilers

Chapter 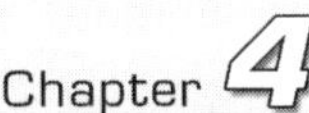

Steam Systems

Section 4.2—Steam Traps and Strainers

Name ______________________________ **Date** ______________________

True-False

T F **1.** A steam strainer should be located in a steam line after a steam trap.

T F **2.** Steam loses heat at the point of use.

T F **3.** Return steam traps discharge condensate directly back to the boiler.

T F **4.** Return steam traps are commonly found in modern boiler plants.

T F **5.** To prevent water hammer, condensate is removed from steam lines by placing steam traps where condensate buildup could occur.

T F **6.** A float thermostatic steam trap is a steam trap that has a bellows filled with a fluid that boils at steam temperature.

Multiple Choice

______________ **1.** Steam traps are ___ devices that increase overall system efficiency by removing air and condensate without the loss of steam.

A. manual
B. electrical
C. automatic
D. semiautomatic

______________ **2.** Steam traps remove air and ___ from steam lines and heating units.

A. condensate
B. oil
C. gas
D. steam

______________ **3.** Condensate in steam lines can result in ___.

A. greater boiler efficiency
B. pure steam production
C. foaming
D. water hammer

__________ **4.** Four common types of ___ include the thermostatic, float thermostatic, inverted bucket, and thermodynamic.
A. return tanks
B. flow indicators
C. steam strainers
D. nonreturn steam traps

__________ **5.** A relatively new method of testing steam traps is to measure ___ to determine the presence of steam.
A. temperature
B. conductivity
C. the NOWL
D. radiation

__________ **6.** Steam strainers should be located on a steam line ___.
A. before a steam trap
B. after a steam trap
C. after a feedwater heater
D. before a steam header

__________ **7.** A(n) ___ is a device used to test steam trap function by comparing sound waves emitted during different load conditions with a properly operating steam trap.
A. infrared thermometer
B. contact indicator
C. sound flow indicator
D. ultrasonic tester

__________ **8.** A thermodynamic steam trap opens and closes by a single ___ that rises to allow the discharge of air and cool condensate.
A. float
B. movable disc
C. electric sensor
D. flexible bellows

__________ **9.** In a float thermostatic trap, the ball float opens and closes a discharge valve depending on the amount of ___ in the chamber.
A. condensate
B. steam
C. feedwater chemicals
D. water and steam

__________ **10.** A steam trap that fails to open causes a heating unit to become ___.
A. steambound
B. waterlogged
C. very hot
D. contaminated with fuel oil

Steam Traps

_______________ **1.** Thermodynamic

_______________ **2.** Inverted bucket

_______________ **3.** Thermostatic

_______________ **4.** Float thermostatic

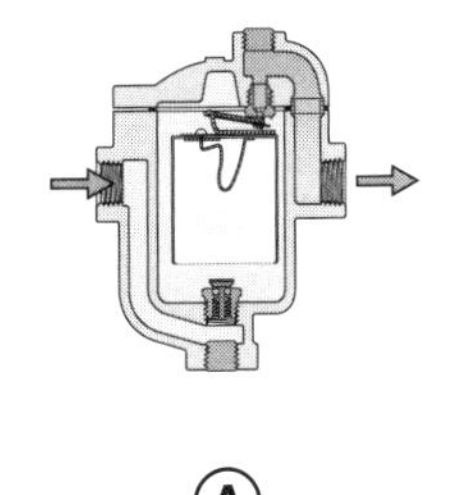

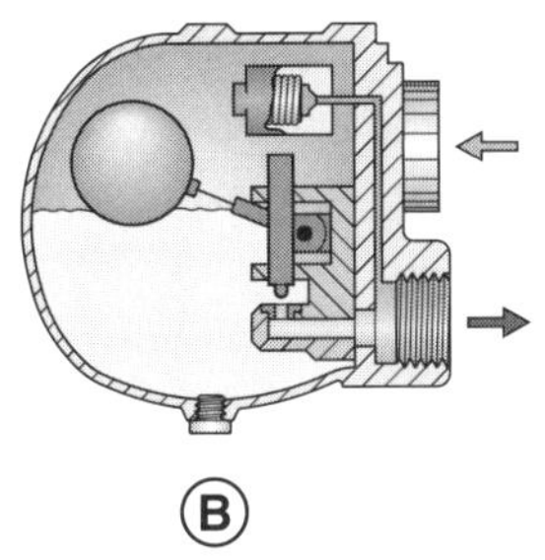

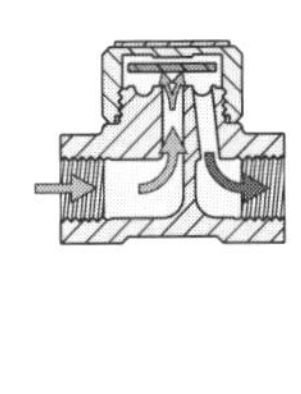

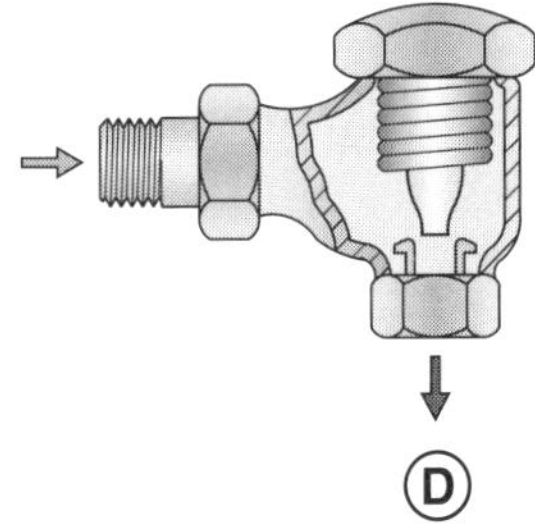

Steam Trap Operation

_______________ **1.** Condensate discharge outlet

_______________ **2.** Inverted bucket

_______________ **3.** Discharge valve

_______________ **4.** Steam and condensate inlet

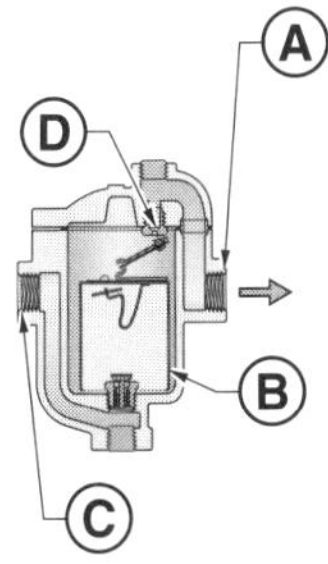

_______________ **5.** Steam and condensate inlet

_______________ **6.** Movable disc closed position

_______________ **7.** Trapped steam in control chamber

_______________ **8.** Condensate discharge outlet

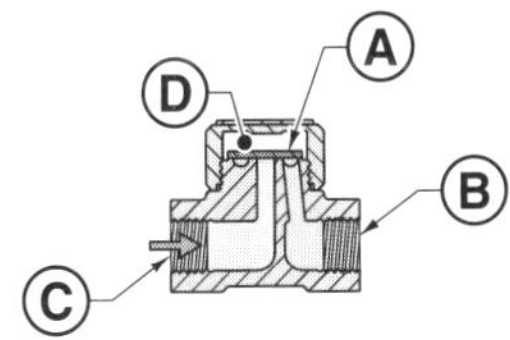

_______________ **9.** Steam and condensate inlet

_______________ **10.** Discharge valve

_______________ **11.** Condensate discharge outlet

_______________ **12.** Bellows

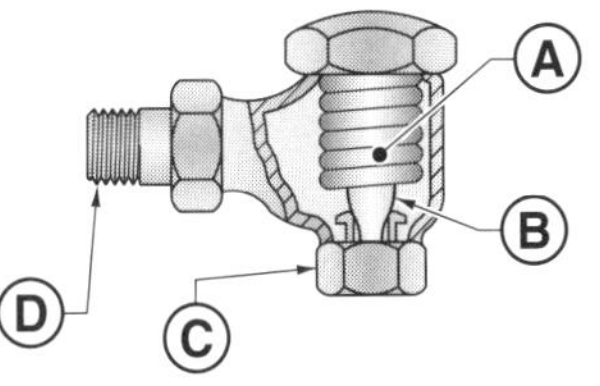

_______________ **13.** Discharge valve

_______________ **14.** Steam and condensate inlet

_______________ **15.** Condensate discharge outlet

_______________ **16.** Ball float

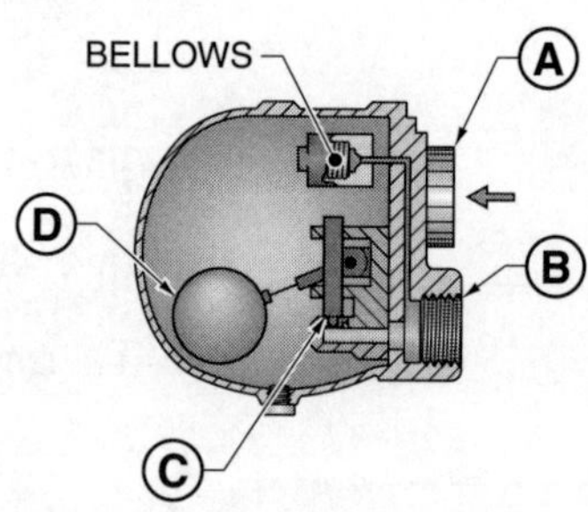

Steam Trap Location

_______________ **1.** Feedwater pump

_______________ **2.** Check valve

_______________ **3.** Main steam line

_______________ **4.** Steam trap

_______________ **5.** Heating unit

_______________ **6.** Stop valve

_______________ **7.** Main steam stop valve

_______________ **8.** Condensate return tank

_______________ **9.** Strainer

_______________ **10.** Main steam header

_______________ **11.** Condensate return line

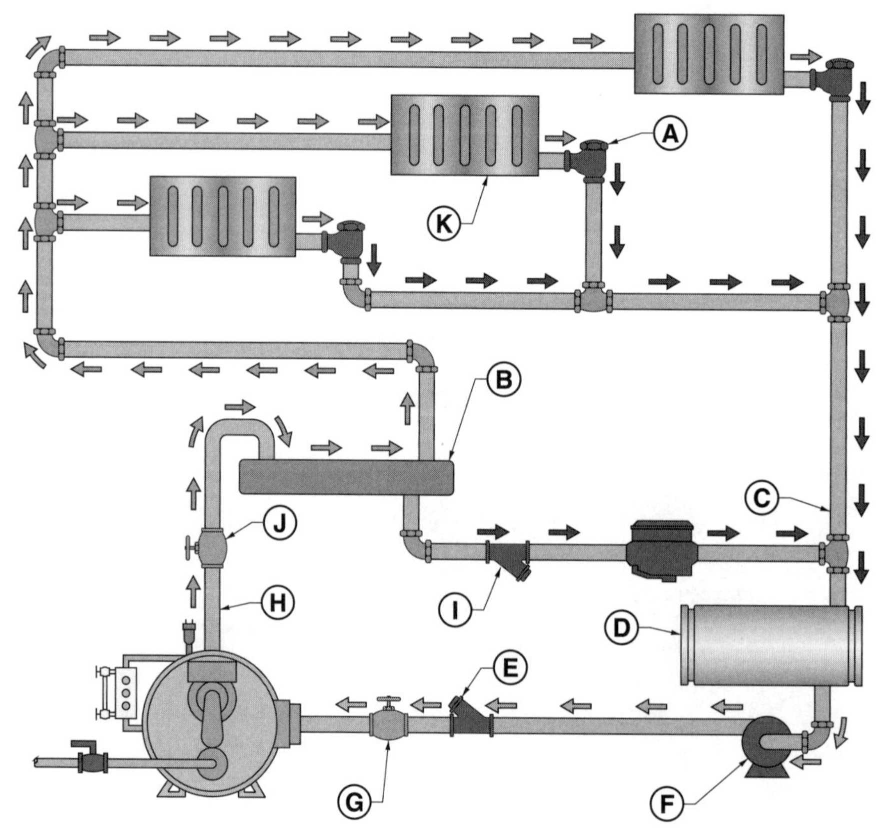

Steam Trap Testing Devices

_______________ **1.** Contact thermometer

_______________ **2.** Temperature-indicating crayon

_______________ **3.** Flow indicator

_______________ **4.** Ultrasonic tester

_______________ **5.** Infrared thermometer

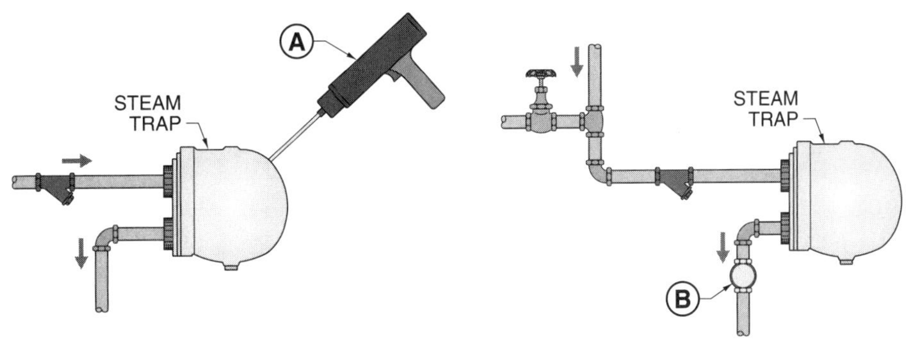

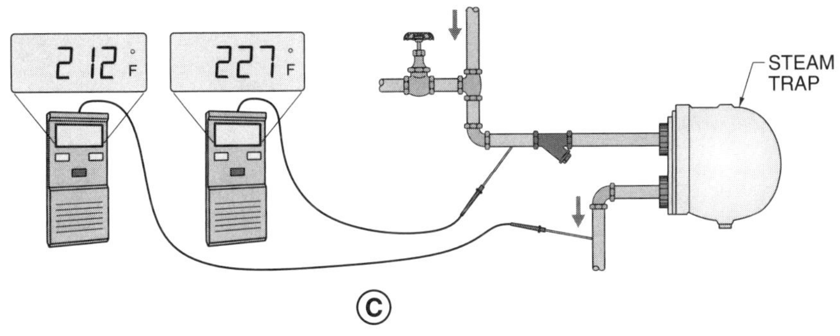

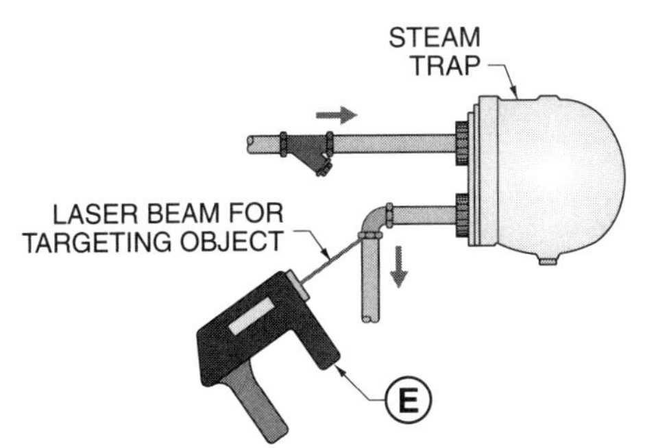

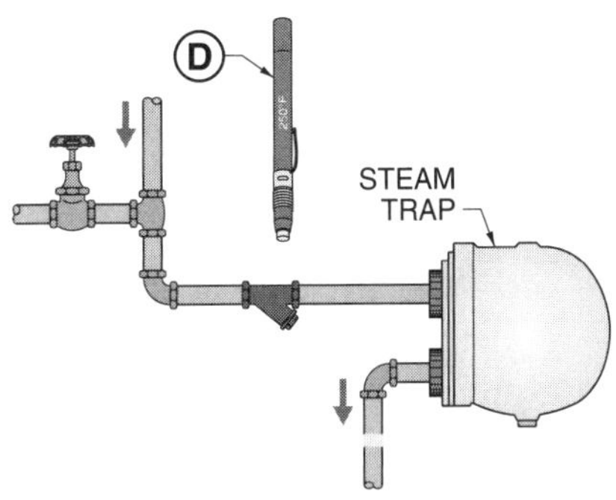

Testing Steam Traps

__________ **1.** Steam trap functioning normally

__________ **2.** Steam trap malfunction

__________ **3.** Strainer clogged

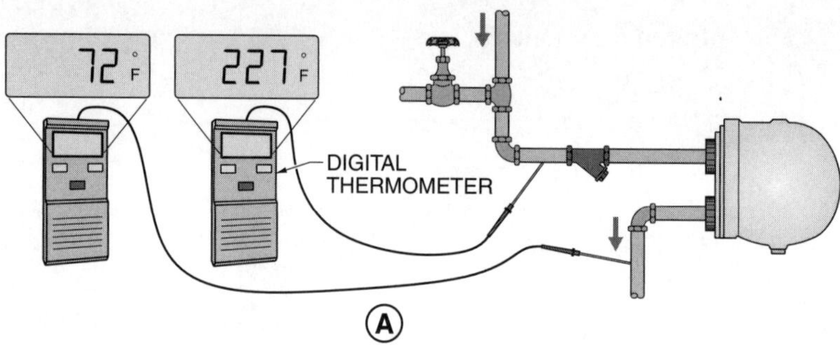

Ⓐ

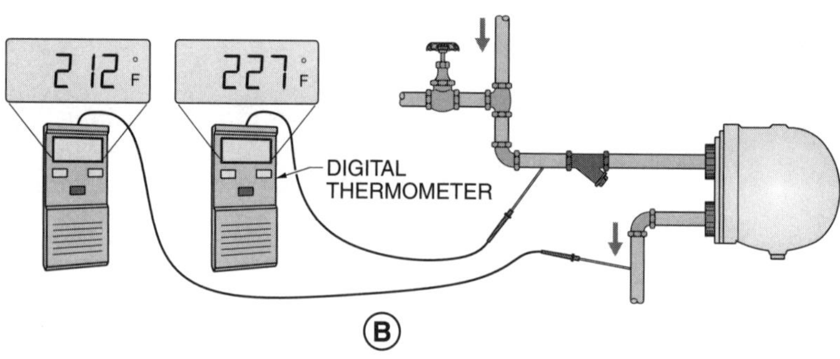

Ⓑ

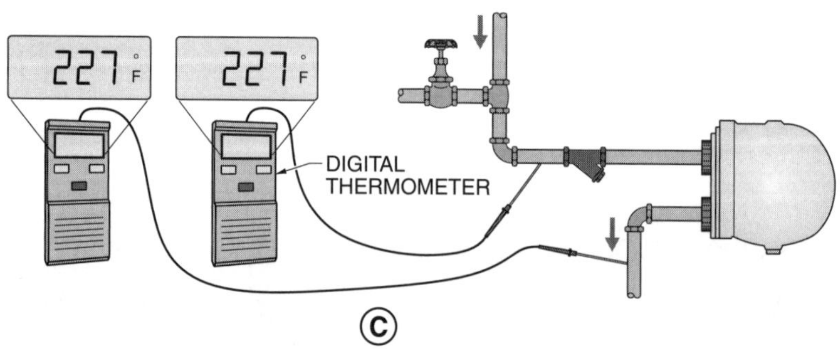

Ⓒ

Chapter 

Fuel Systems

Section 5.1—Gas Systems

Name ______________________ **Date** ______________

True-False

T F **1.** The most common LP gas is propane.

T F **2.** In a low pressure gas system, the manual reset valve cannot open until the gas pilot is lit.

T F **3.** During the combustion process, fuel mixed with air is burned to produce the heat necessary to operate a boiler.

T F **4.** In a low pressure gas system, gas pressure is reduced from supplier gas pressure to approximately 3 psi.

T F **5.** Natural gas is a colorless and odorless fossil fuel.

T F **6.** In a high pressure gas system, gas is supplied through a high pressure gas main at a higher pressure than that within a typical residential gas supply line.

T F **7.** A therm is the quantity of gas required to produce 100,000 Btu.

T F **8.** Gas burners can only be high pressure gas burners.

Multiple Choice

______________ **1.** In a low pressure gas burner, gas is mixed with air in the ___.

A. burner
B. mixing chamber before the burner
C. combustion chamber
D. boiler furnace

______________ **2.** In a high pressure gas burner, the gas mixes with the air on the inside of the ___.

A. burner
B. mixing chamber
C. combustion chamber
D. boiler furnace

_______________ **3.** A ___ is a narrowed portion of a tube.
A. solenoid
B. butterfly valve
C. venturi
D. pilot regulator

_______________ **4.** The heat content of natural gas varies from 950 Btu to 1050 ___.
A. pounds per square inch
B. calories per cubic inch
C. therms
D. Btu per cubic foot

_______________ **5.** In a high pressure gas system, the plant pressure ___ valve reduces gas pressure to line pressure used in the system.
A. butterfly
B. regulating
C. safety
D. pilot

Low Pressure Gas Systems

_______________ **1.** Butterfly valve

_______________ **2.** Gas pressure switch

_______________ **3.** Gas burner

_______________ **4.** Main gas shutoff cock

_______________ **5.** Pilot shutoff cock

_______________ **6.** Blower

_______________ **7.** Gas pilot

_______________ **8.** Manual reset valve

_______________ **9.** Mixing chamber

_______________ **10.** Gas supply line

_______________ **11.** Main gas solenoid valve

_______________ **12.** Pilot solenoid valve

_______________ **13.** Venturi

_______________ **14.** Gas pressure regulator

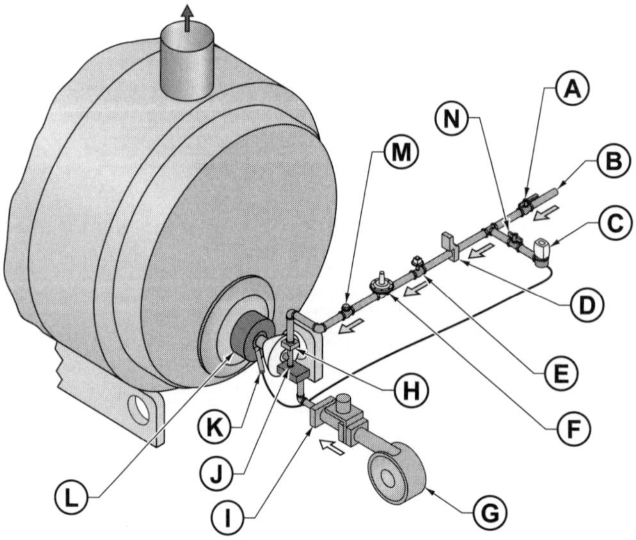

High Pressure Gas Systems

______	**1.** Utility meter	______	**7.** Butterfly gas valve
______	**2.** Pilot vent valve	______	**8.** Pilot pressure regulator
______	**3.** Main gas vent valve	______	**9.** Pilot valves
______	**4.** Pilot adjusting cock	______	**10.** Pilot pressure gauge
______	**5.** Utility pressure regulating valve	______	**11.** Main gas shutoff cock
______	**6.** Low gas pressure switch	______	**12.** Pilot shutoff cock
		______	**13.** Main gas valve
		______	**14.** Plant pressure regulating valve

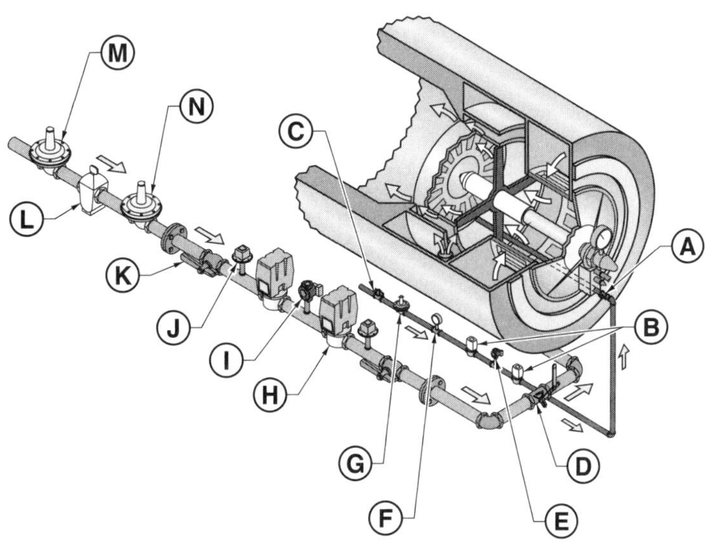

Chapter 

Fuel Systems

Section 5.2—Fuel Oil Systems

Name ______________________________ **Date** ______________

True-False

T F **1.** The fire point temperature is higher than the flash point temperature.

T F **2.** Fuel oil characteristics include viscosity, pour point, flash point, fire point, volatility, and heating value.

T F **3.** Atomization is the ability of fuel to be vaporized.

T F **4.** The Btu rating of a fuel indicates how much heat it can produce.

T F **5.** A decrease in a suction gauge reading indicates either a dirty strainer or cold fuel oil.

T F **6.** Fuel oil can be classified as distillate fuel oil or heavy fuel oil.

T F **7.** Heavy fuel oils are relatively easy to pump and atomize for burning.

T F **8.** Rotary cup burners can burn only No. 2 fuel oil.

T F **9.** Rotary cup burners atomize fuel oil using a spinning cup and high-velocity air.

T F **10.** Air used to atomize fuel oil is primary air.

T F **11.** If fuel oil is not properly blended, it may stratify in a tank.

T F **12.** In an air atomizing burner, steam and fuel are discharged to the burner for combustion after the flames reach the end of the combustion chamber.

T F **13.** Fuel oil atomization temperature varies with the fuel oil grade and burner.

T F **14.** A safety valve prevents backflow to a fuel tank from a fuel oil system.

T F **15.** A modulating cam is a mechanical device used to regulate the flow of fuel oil to a burner for controlling the firing rate.

T F **16.** The most common fuel oil strainers are the T strainer and basket strainer.

T F **17.** A therm is equivalent to 100,000 Btu.

T F **18.** Fuel oil is a liquid fossil fuel that consists primarily of carbon, hydrogen, and moisture.

T F **19.** A high vacuum of 10″ or more on a suction pressure gauge indicates either cold fuel oil or dirty suction strainers.

T F **20.** Fuel oil burners include atomizing and rotary cup burners.

T F **21.** Condensate may be directed to a condensate return tank or wasted to prevent boiler water contamination.

T F **22.** The combination gas/fuel oil burner has in integral gas burner that allows quick changeover from one fuel to another.

Multiple Choice

______________ **1.** When burning No. 6 fuel oil, the fuel oil strainers should be cleaned at least once every ___ hr.
- A. 8
- B. 10
- C. 12
- D. 24

______________ **2.** When replacing the flange cover in a duplex fuel oil strainer, the ___ must be carefully replaced to prevent air from entering the system.
- A. plug valve
- B. gasket
- C. hand screw
- D. basket

______________ **3.** A ___ pump draws fuel oil from a fuel oil tank and delivers it to a burner at a controlled pressure.
- A. transfer
- B. fuel oil
- C. condensate
- D. circulating

______________ **4.** A ___ valve protects a fuel oil pump from excessive pressure by discharging fuel oil through the return line to the fuel oil tank.
- A. safety
- B. bypass
- C. relief
- D. stop

______________ **5.** ___ regulations place restrictions on the design and installation of fuel oil tanks.
- A. EPA
- B. OSHA
- C. ASME
- D. NIOSH

______ **6.** Fuel oil burners are designed to provide fuel oil in a ___.
A. steady stream
B. fine spray
C. half spray, half stream
D. pulsing stream

______ **7.** A ___ is a burner accessory that opens when energized by contacts in the programmer to allow fuel oil to flow to a burner nozzle.
A. fuel oil pressure regulator
B. fuel oil controller
C. main fuel oil solenoid valve
D. metering valve

______ **8.** The ___ of fuel oil is the lowest temperature at which it will flow as a liquid.
A. fire point
B. flash point
C. pour point
D. viscosity

______ **9.** The ___ of fuel oil is the temperature at which fuel oil gives off vapor that flashes when exposed to an open flame.
A. fire point
B. flash point
C. pour point
D. viscosity

______ **10.** The ___ is the temperature at which fuel oil will burn continuously when exposed to an open flame.
A. fire point
B. flash point
C. pour point
D. viscosity

______ **11.** The internal resistance of fuel oil to flow is the ___.
A. fire point
B. flash point
C. pour point
D. viscosity

______ **12.** In order to reduce the viscosity of fuel oil, it is necessary to ___.
A. decrease its temperature
B. decrease its pour point
C. increase its temperature
D. decrease its pressure

______________ **13.** ___ is the process of reducing fuel oil into a fine spray of minute particles.
A. Ultimate analysis
B. Flue-gas analysis
C. High-volatile testing
D. Atomization

______________ **14.** A(n) ___ burner is a fuel oil burner that uses air, steam, or fuel oil pressure to atomize fuel oil for maximum combustion efficiency.
A. vaporizing
B. combustion
C. rotary cup
D. atomizing

______________ **15.** ___ the fuel oil is the process of using air to discharge residual fuel oil in a burner nozzle.
A. Atomizing
B. Purging
C. Venting
D. Vaporizing

Fuel Oil Pumps

______________ **1.** Crescent seal

______________ **2.** Shaft

______________ **3.** Suction side

______________ **4.** Drive gear

______________ **5.** Ring gear

______________ **6.** Discharge side

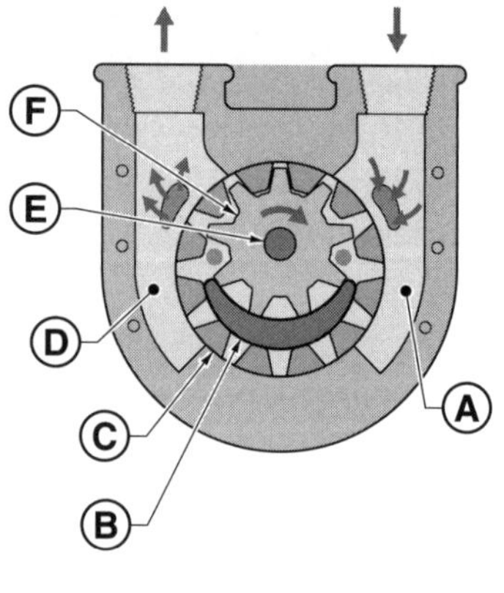

F
E
D
C
B
A

Fuel Oil Burners

_______________ **1.** Rotary cup

_______________ **2.** Air atomizing

_______________ **3.** Pressure atomizing

_______________ **4.** Steam atomizing

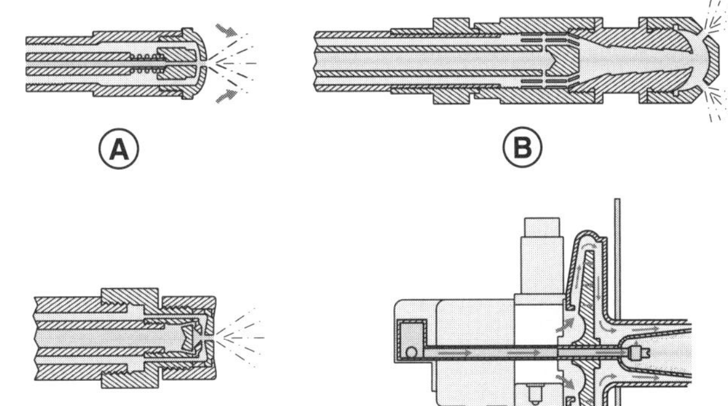

Fuel Oil Systems

_______________ **1.** Vacuum gauge

_______________ **2.** Fuel oil relief valve

_______________ **3.** Nozzle air pressure gauge

_______________ **4.** Fuel oil controller

_______________ **5.** Fuel oil strainer

_______________ **6.** Modulating cam

_______________ **7.** Fuel oil burner pressure gauge

_______________ **8.** Check valve

_______________ **9.** Fuel oil thermometer

_______________ **10.** Metering valve

_______________ **11.** Main fuel oil solenoid valves

_______________ **12.** Fuel oil pressure gauge

_______________ **13.** Fuel oil pump

_______________ **14.** Fuel oil pressure regulator

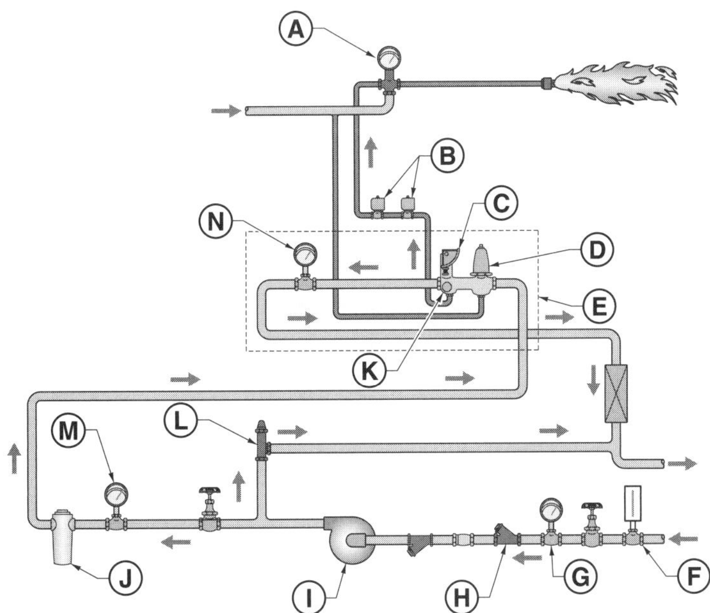

Fuel Oil Strainer

_______________ **1.** Strainer basket

_______________ **2.** Basket housing

_______________ **3.** Hand screw

_______________ **4.** Drain plug

_______________ **5.** Plug valve

_______________ **6.** Handle position indicates basket in service

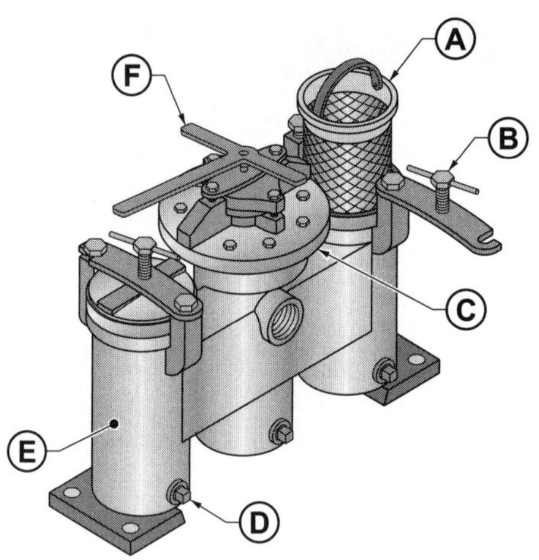

Fuel Systems

Section 5.3—Combustion Controls

Name ______________________________ Date ______________

True-False

T F 1. Complete combustion is combustion that occurs when all fuel is burned with a minimum amount of excess air remaining.

T F 2. If combustion is not complete before gases come in contact with a heating surface, the gases will cool and create soot and smoke.

T F 3. A flame scanner proves the main flame only.

T F 4. An operating pressure control starts and shuts down a burner based on steam pressure.

T F 5. A flame establishing period regulates burner firing rate by controlling a modulating motor connected to a combustion equipment control linkage.

T F 6. A turndown ratio must be regulated because if there is an increase in the fuel supplied, then less primary and secondary air will be required.

T F 7. Run period is the period after ignition trials but before the operating set point is reached while the main burner is firing.

T F 8. A boiler management and control system may also be known as a boiler control system (BCS) or microcomputer boiler control system (MBCS).

T F 9. Combustion efficiency can be determined by analyzing flue gas.

T F 10. A flame scanner sensing device that contains lead sulfide is used to detect infrared light.

Multiple Choice

______________ 1. A(n) ___ is an air pressure-activated switch that closes after proving sufficient pressure of combustion air from the forced draft fan.

A. air proving switch
B. pilot modulating relay
C. combustion flowmeter
D. pressure switch flowmeter

_______________ **2.** A(n) ___ sensor is a flame sensor that senses light frequencies that are higher than those visible to the eye.
A. photocell
B. infrared
C. ultraviolet
D. pilot

_______________ **3.** A(n) ___ is a safety device that senses if the pilot light and/or main flame are lit.
A. low water fuel cutoff
B. annunciator
C. flame scanner
D. heater thermostat

_______________ **4.** A(n) ___ system is a solid-state control system in which a building automation controller is wired directly to control devices.
A. ultraviolet burner control (UBC)
B. modulating control status (MCS)
C. programmed emission safeguard (PES)
D. direct digital control (DDC)

_______________ **5.** A(n) ___ system is a combustion control system that controls the amount of steam produced by starting and stopping the boiler.
A. modulating control
B. ON/OFF control
C. burner proving
D. power control

_______________ **6.** ___ combustion is combustion that occurs when all fuel is not burned.
A. Complete
B. Incomplete
C. Perfect
D. Imperfect

_______________ **7.** A(n) ___ is a combustion control system that controls the amount of steam produced by changing the burner firing rate.
A. ON/OFF control system
B. aquastat
C. temperature flowmeter
D. modulating control system

_______________ **8.** A ___ is burner control equipment that monitors the burner start-up sequence and the main flame during normal operation.
A. flame safeguard system
B. flue gas analyzer
C. modulation control system
D. flame scanner

______________ 9. A ___ is a control that functions as the mastermind of a burner control system to control the firing cycle.
A. pressure control
B. butterfly solenoid
C. programmer
D. primary combustion analyzer

______________ 10. ___ is the amount of fuel a burner is capable of burning in a given unit of time.
A. High fire
B. Low fire
C. Firing rate
D. Heating surface

______________ 11. ___ fire is burning the maximum amount of fuel in a given unit of time.
A. Low
B. High
C. Minimum
D. Medium

______________ 12. ___ air controls the amount of fuel oil capable of being burned.
A. Forced
B. Primary
C. Secondary
D. Tertiary

______________ 13. ___ air controls combustion efficiency and is usually introduced into a furnace from below the burner.
A. Forced
B. Primary
C. Secondary
D. Tertiary

______________ 14. When a flame safeguard programmer sequences burner function, ___ is the period of time during which the pilot and main burner must be lit.
A. prepurge
B. pilot flame-establishing period
C. ignition trials
D. postpurge

______________ 15. A(n) ___ is a boiler condition that occurs when the flame in a boiler furnace has been unintentionally lost.
A. meltdown
B. flame failure
C. ultimate analysis
D. shutdown

_______________ **16.** Unburned fuel oil that is heated in a furnace will ___.
A. turn into steam
B. vaporize
C. solidify
D. flow slowly

_______________ **17.** A photocell sensor senses ___ light.
A. near infrared
B. ultraviolet
C. visible
D. far infrared

_______________ **18.** If a flame scanner does not prove a pilot flame during the pilot flame-establishing period, the pilot ___ will be de-energized, preventing fuel flow to the burner.
A. solenoid valve
B. flames sensor
C. high fire valve
D. fuel strainer

_______________ **19.** Well-designed burners firing gaseous and liquid fuels operate at excess air levels of about ___% and result in negligible amounts of unburned fuel.
A. 0.433
B. 15.0
C. 25.0
D. 212.0

_______________ **20.** The ___ of a modulating burner is the ratio of the maximum firing rate to the minimum firing rate.
A. firing rate
B. burning ratio
C. maximizer rate
D. turndown ratio

_______________ **21.** ___ is the condition where a flame travels upwind and into the burner assembly.
A. Blowdown
B. Blowback
C. Flashdown
D. Flashback

_______________ **22.** The primary function of a boiler management and control system is ___.
A. low water protection
B. combustion safety
C. controlling high water
D. combustion efficiency

______________ **23.** When an increase in steam pressure is required, the ___ activates the programmer to start a purge cycle and lights the pilot light.

A. aquastat
B. remote sensor
C. pressure control
D. pressure gauge

______________ **24.** ___ combustion occurs when all fuel burns using only the theoretical amount of air.

A. Imperfect
B. Perfect
C. Complete
D. Incomplete

______________ **25.** ___ prevents blower operation at startup until the damper reaches purge position.

A. An energy-saving prepurge
B. Energy-saving intelligence
C. Remote monitoring
D. An ON/OFF control

Modulating Control

______________ **1.** Bellows

______________ **2.** Pressure setting

______________ **3.** To siphon

______________ **4.** Differential setting

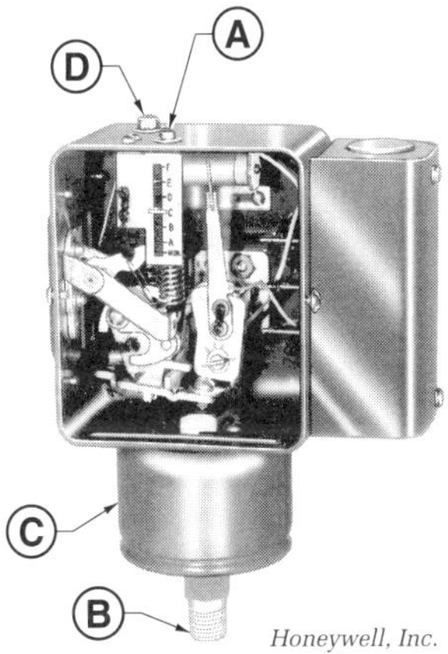

A
D
C
B

Honeywell, Inc.

Flame Sensors

_______________ **1.** Ultraviolet

_______________ **2.** Flame rod

_______________ **3.** Infrared

_______________ **4.** Photocell

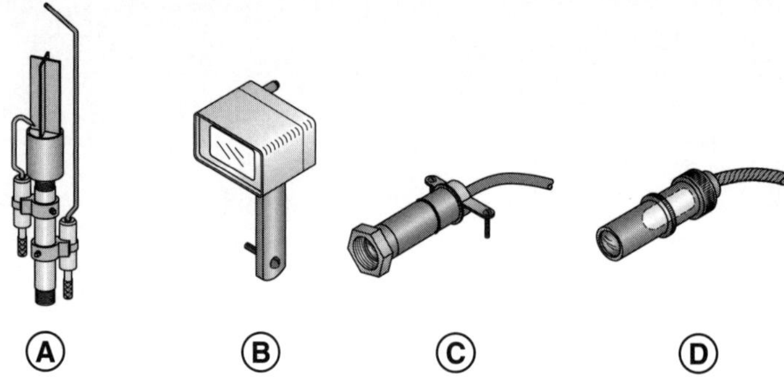

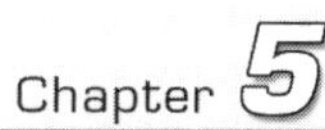

Fuel Systems

Section 5.4—Air Pollution

Name ______________________________ **Date** ________________

True-False

T F **1.** Air pollutants are classified as primary pollutants and secondary pollutants.

T F **2.** Air pollution regulations are enacted at the federal level and/or at the state and local level.

T F **3.** A primary pollutant is a pollutant formed from the interaction between two or more secondary pollutants.

T F **4.** The EPA (NAAQS) set ambient pollutant standards to address 12 criteria pollutants.

T F **5.** The most recent amendments to the Clean Air Act occurred in 1997.

T F **6.** A state implementation plan is a plan that gives the states the responsibility for developing their own programs to reduce air pollution.

T F **7.** A major source facility is a facility that emits less than 10 tons per year of any single air toxic and less than 25 tons per year of any combination of air toxics.

T F **8.** Area source boilers emit only small amounts of pollutants, primarily because most burn natural gas.

Multiple Choice

______________ **1.** A(n) ___ is matter that contaminates air, water, or soil.

A. particulate
B. discharge
C. emission
D. pollutant

______________ **2.** ___ are pollution standards for six priority pollutants set by the Environmental Protection Agency through the Clean Air Act.

A. PLCs
B. BMCS
C. EPAQS
D. NAAQS

_______________ **3.** The 1990 Clean Air Act Amendment is comprised of ___ titles that have the potential to affect nearly every source of air pollution.
A. 7
B. 9
C. 11
D. 13

_______________ **4.** The ___ are environmental standards for ambient pollutant emissions from new sources of emissions.
A. NSPS
B. NAAQS
C. CCA
D. SIP

_______________ **5.** A(n) ___ facility is a facility that emits 10 tons or more per year of any single air toxic or 25 tons or more per year of any combination of air toxics.
A. area source
B. major source
C. minor source
D. emissions

Chapter 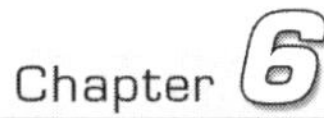

Draft Systems

Section 6.1—Draft

Name ______________________________ Date ________________

True-False

T F **1.** Draft is the flow of air or gases of combustion caused by a difference in pressure between two points.

T F **2.** Air or gases flow from a location of low pressure to a location of high pressure.

T F **3.** The two types of draft produced in a boiler are mechanical and forced.

T F **4.** Mechanical draft is produced by a fan or blower.

T F **5.** Natural draft can be regulated by controlling the temperature of the gases of combustion.

T F **6.** Dampers regulate draft by restricting the flow of air and/or the gases of combustion.

T F **7.** In the combustion process, approximately 75 lb of air are required for every pound of fuel burned.

T F **8.** The amount of natural draft produced increases with the height of a stack.

T F **9.** The types of mechanical draft are forced, induced, and a combination of forced and induced.

T F **10.** An induced draft fan is located in front of a boiler furnace.

T F **11.** A variable-speed drive (VSD) is a motor controller used to vary the frequency of the electrical signal supplied to an AC motor in order to control its rotational speed.

T F **12.** Balanced draft is a mechanical draft from fans located before and after a boiler furnace.

T F **13.** A forced draft fan is located in the breeching.

T F **14.** Draft is typically measured in pounds per square inch (psi).

T F **15.** Natural draft produces greater amounts of draft in the winter than in the summer.

T F **16.** When using a manometer to measure draft, the difference between the levels of liquid in two legs is the draft measurement.

T F **17.** Larger boilers are designed to use mechanical draft, which can be controlled more precisely and results in higher combustion efficiency.

T F **18.** The amount of draft determines the rate of combustion.

Multiple Choice

______ **1.** A draft system is a boiler system that regulates the ___.
A. analysis of the gases of combustion
B. flow of air to and from the burner
C. electrical signal supplied to an AC motor
D. stack condensation

______ **2.** Natural draft is regulated using ___ to control the flow of gases of combustion into a stack.
A. draft fans
B. gas pressure regulators
C. regulating valves
D. dampers

______ **3.** Air flow in a natural draft system is controlled as necessary through the ___.
A. boiler and breeching
B. breeching and induced draft fan
C. boiler drum and furnace
D. forced draft fan and inlet damper

______ **4.** A furnace can be made hotter by ___.
A. decreasing the amount of the air-fuel mixture being burned
B. increasing the amount of the air-fuel mixture being burned
C. decreasing the flow rate of hot gases
D. increasing the stack height

______ **5.** A ___ can save energy because it can precisely control a fan's motor speed so that the fan volume more efficiently matches the boiler air flow requirements.
A. draft system
B. power-driven fan
C. VSD
D. damper

______ **6.** ___ draft is air that is pulled through a boiler furnace with fans located in the breeching.
A. Induced
B. Forced
C. Unforced
D. Natural

______ **7.** A negative pressure reading in a manometer is indicated if the level of the liquid in the leg connected to the breeching is ___ to the atmosphere.
A. higher than the level in the leg left open
B. lower than the level in the leg left open
C. higher than the level in the leg left closed
D. lower than the level in the leg left closed

______________ **8.** A ___ is a device used to create resistance in order to regulate the flow of air and the gases of combustion in a boiler.
A. regulator
B. variable-speed drive
C. fan
D. damper

______________ **9.** The scales used on diaphragm gauges vary from hundredths of an inch to several inches of ___, depending on the pressure being measured.
A. mercury
B. water column
C. saturated steam
D. flow

______________ **10.** A ___ is a device that measures draft by comparing pressures at two locations with a U-tube or inclined tube gauge.
A. pyrometer
B. dynometer
C. manometer
D. trynometer

Manometer Readings

______________ **1.** Positive reading

______________ **2.** No reading

______________ **3.** Negative reading

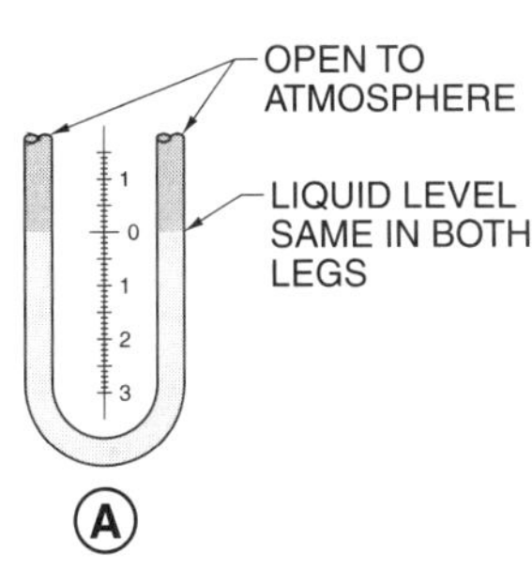

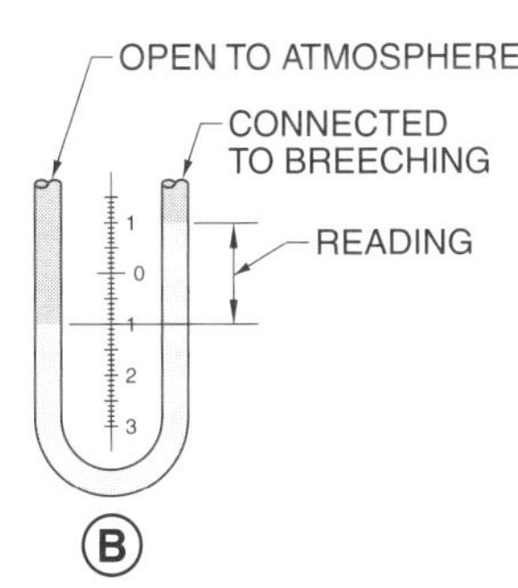

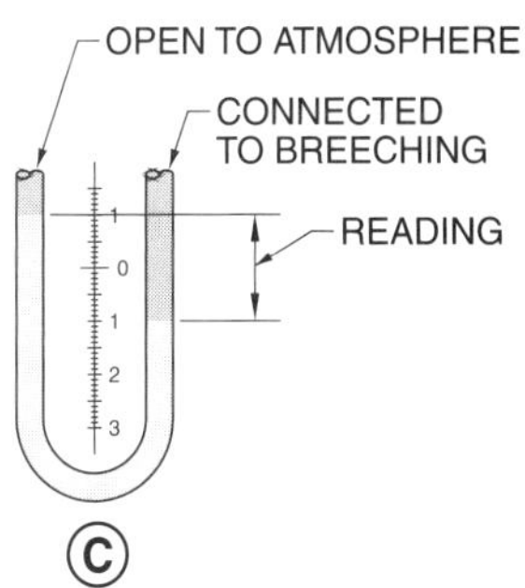

Natural Draft

______________ 1. Cold air for combustion

______________ 2. Hot gases of combustion

______________ 3. Gases of combustion released to atmosphere

______________ 4. Cold ambient air

______________ 5. Stack

______________ 6. Heat

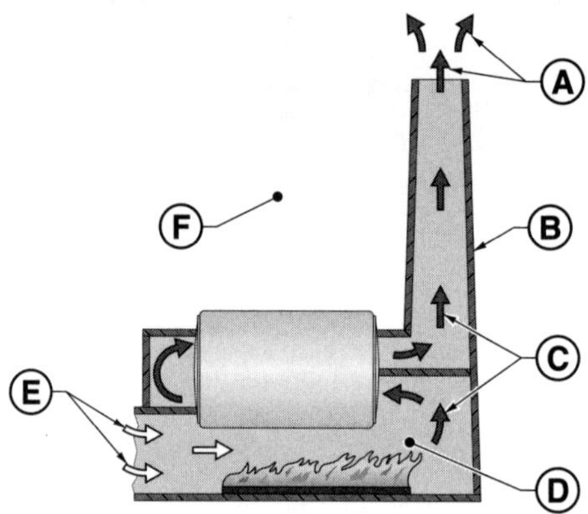

Forced Draft

______________ 1. Forced draft fan

______________ 2. Stack

______________ 3. Gases of combustion

______________ 4. Air flow

______________ 5. Boiler drum

______________ 6. Outlet damper

______________ 7. Furnace

______________ 8. Air entering furnace

______________ 9. Inlet damper

______________ 10. Breeching

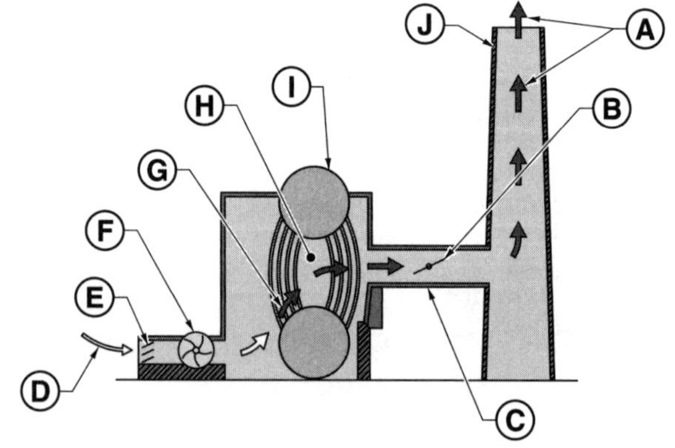

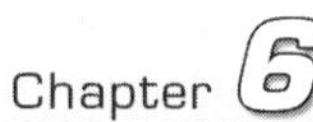

Chapter 6

Draft Systems

Section 6.2—Draft Control

Name ______________________________ **Date** ____________________

True-False

T F **1.** The method for controlling draft is determined by the boiler equipment.

T F **2.** Adjusting draft controls without proper training and equipment can cause poor air flow through a boiler.

T F **3.** Proper control of draft maintains low combustion efficiency.

T F **4.** Properly burned fuel always results in carbon particles, soot, or excessive smoke.

T F **5.** The words "stack" and "chimney" are often used interchangeably when referring to the structure used to direct the flow of gases of combustion from a boiler to the atmosphere.

Multiple Choice

______________ **1.** High-efficiency condensing boilers may use steel or ___ stacks because of the low stack temperatures in these types of boilers.

A. aluminum
B. PVC
C. metal
D. plastic

______________ **2.** If the stack height is over ___′ or if an extremely large breeching and stack combination causes excessive draft, a damper can be located in the breeching close to the stack.

A. 10
B. 50
C. 100
D. 150

______________ **3.** With high-efficiency condensing boilers, the stack temperature may be as low as ___°F.

A. 32
B. 100
C. 500
D. 1000

Draft Control

____________ **1.** Burner nozzle

____________ **2.** Blower motor

____________ **3.** Gases of combustion

____________ **4.** Outlet

____________ **5.** Forced draft fan

____________ **6.** Boiler

____________ **7.** Gas pilot assembly

____________ **8.** Intake

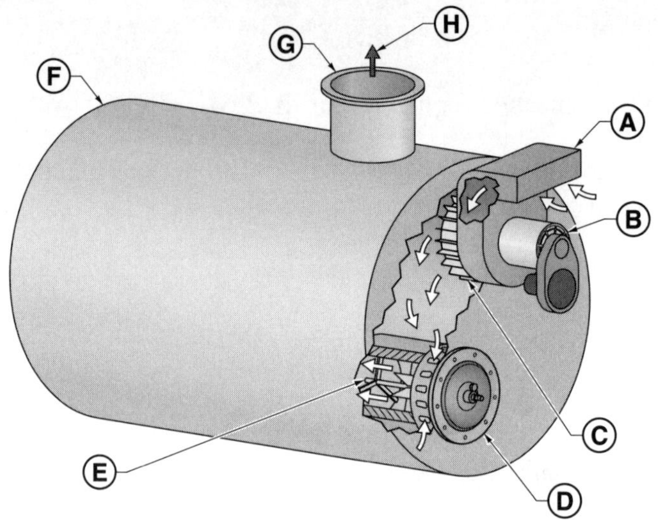

Stacks

____________ **1.** Stack

____________ **2.** Boiler

____________ **3.** Drain connection

____________ **4.** Cleanout

____________ **5.** Forced draft fan

____________ **6.** Breeching

____________ **7.** Gases of combustion to atmosphere

____________ **8.** Outlet

____________ **9.** Offset stack

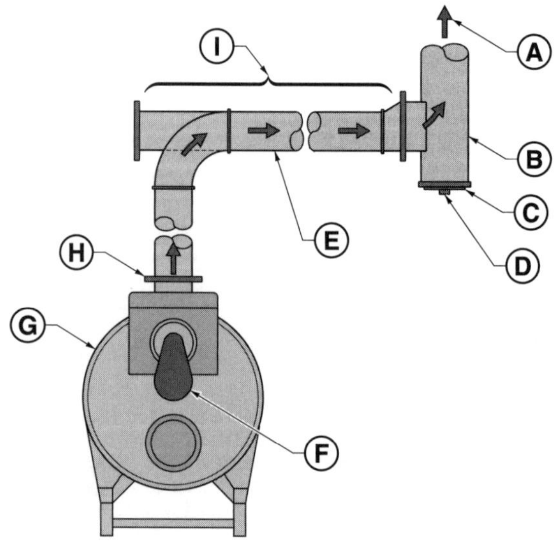

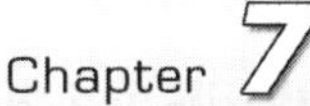

Chapter 7

Boiler Water Treatment

Section 7.1—Boiler Water Conditions

Name ______________________________ **Date** ________________

True-False

T F **1.** Dissolved solids are impurities such as calcium, silica, and iron dissolved in solution.

T F **2.** Priming and carryover or a failed steam trap can lead to water hammer and possible pipe rupture.

T F **3.** Oxygen in boiler water is used to remove nonadhering sludge.

T F **4.** Water containing dissolved minerals is soft water.

T F **5.** Scale acts as an insulating material and reduces the transfer of heat to water.

T F **6.** Chemical treatment changes calcium and magnesium into nonadhering sludge to prevent scale buildup.

T F **7.** Ground water is water that is treated for use in a boiler.

T F **8.** A boil-out procedure is performed when a new boiler is installed or repaired.

T F **9.** A boiler must be taken out of service if there are any signs of metal breakdown.

T F **10.** Priming and carryover can be caused by opening a main steam valve too quickly.

T F **11.** Corrosion is the accumulation of sludge on boiler heating surfaces.

T F **12.** High surface tension is a boiler water condition in which the surface of the water is strengthened by a layer of impurities.

T F **13.** Scale and sediment cause pitting of boiler metal surfaces.

T F **14.** Sodium sulfite is an oxygen scavenger that is commonly used to treat boiler water.

T F **15.** Rusting of boiler metal is caused by oxygen in boiler water.

T F **16.** All returns from a fuel oil heater should be properly dumped to eliminate the possibility of further contamination.

T F **17.** Scale formation can result in an increase in fuel required to generate steam.

T F **18.** Foaming causes corrosion of heat transfer surfaces.

Multiple Choice

__________ **1.** Water that contains large quantities of dissolved minerals is called ___.
A. soft water
B. hard water
C. sludge
D. sediment

__________ **2.** Small particles of water carried into steam lines are called ___.
A. scale formation
B. blistering
C. carryover
D. priming

__________ **3.** ___ is a condition caused when steam bubbles are trapped below the boiler water surface.
A. Sludge slugging
B. Foaming
C. Priming
D. Tube corrosion

__________ **4.** ___ is the exposure to highly alkaline elements, causing boiler metal corrosion at stress zones.
A. Sodium sulfate
B. Caustic embrittlement
C. Corrosion
D. Pitting

__________ **5.** ___ combines with oxygen to form sodium sulfate, which accumulates at the bottom of a boiler.
A. Sodium sulfite
B. Zeolite
C. Silica
D. Magnesium

__________ **6.** Nonadhering sludge is best removed when a boiler is ___.
A. under light load
B. under heavy load
C. being drained
D. being tested

__________ **7.** ___ can result in bags and blisters.
A. Overheated breeching
B. An overheated stack
C. Overheated fuel oil
D. Overheated heating surfaces

______________ **8.** Scale is caused by ___.
A. nitrogen
B. carbon monoxide
C. carbon dioxide
D. hard water

______________ **9.** ___ in boiler water causes corrosion and pitting of the boiler metal.
A. Sodium sulfite
B. Sodium sulfate
C. Carbon monoxide
D. Oxygen

______________ **10.** High surface tension on boiler water is shown by ___.
A. pressure above 15 psi
B. steam released by the top try cock
C. increased amounts of sludge
D. fluctuation in the gauge glass

______________ **11.** A ___ is a bulge or deformed area on a boiler shell that extends through the entire thickness and is caused by localized overheating.
A. bag
B. blister
C. boil
D. bubble

______________ **12.** Heating feedwater removes ___.
A. corrosion and scale
B. oxygen and carbon dioxide
C. water hammer
D. silica

Sludge Removal

______________ **1.** To blowdown line

______________ **2.** NOWL

______________ **3.** Bottom blowdown valve

______________ **4.** Boiler water circulation reduced

______________ **5.** Sludge

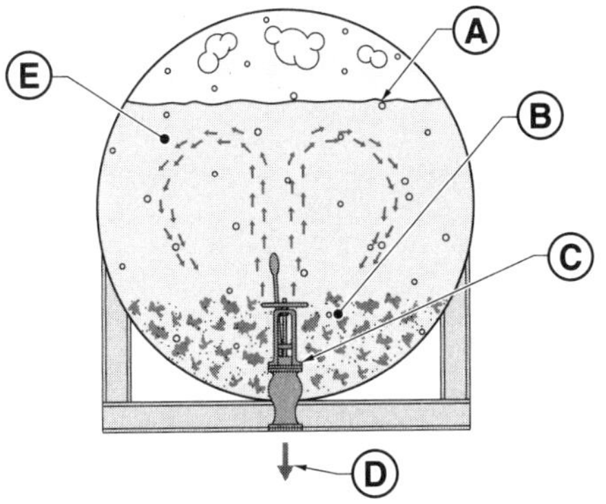

A
B
C
D
E

Priming and Carryover

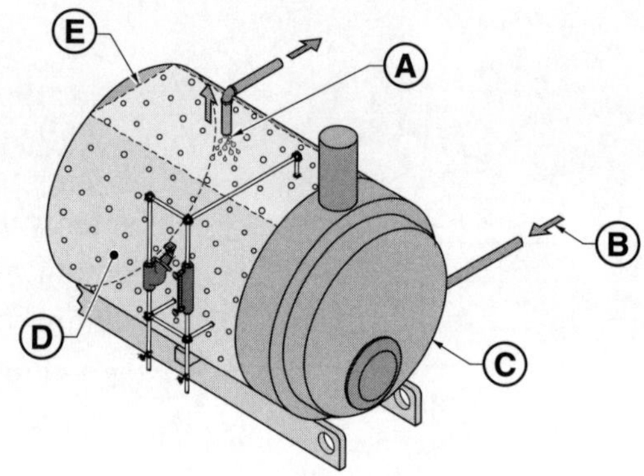

_______________ 1. Boiling water

_______________ 2. Water carried into steam lines

_______________ 3. Feedwater

_______________ 4. High water level

_______________ 5. Boiler

Foaming

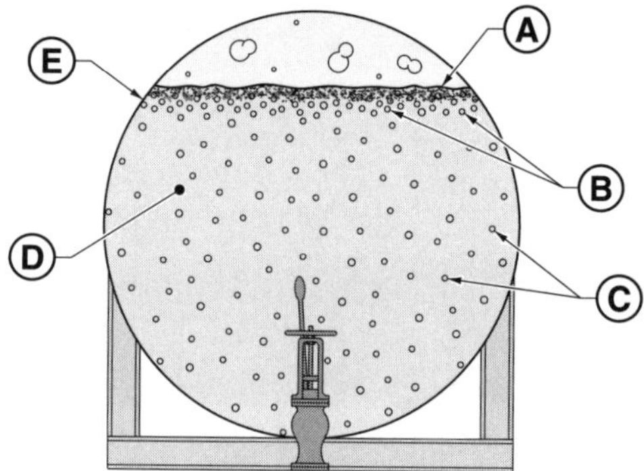

_______________ 1. Trapped steam bubbles

_______________ 2. Boiler

_______________ 3. Boiler water

_______________ 4. Film from impurities

_______________ 5. Steam bubbles

Chapter 7

Boiler Water Treatment

Section 7.2—Boiler Water Analysis

Name ______________________________ **Date** ____________________

True-False

T F **1.** Boiler water analysis tests the condition of boiler water and is used to determine the feedwater treatment required.

T F **2.** A solution with a pH of 0 is neutral.

T F **3.** Either alkaline or acidic water can lead to a boiler explosion.

T F **4.** Water hardness is commonly checked with an electronic hardness tester.

T F **5.** A titration test is a test used in a water treatment test to indicate the presence of a specific substance.

T F **6.** A quicker but less accurate method of determining water hardness is with a soap test.

T F **7.** The amount of solids in water affects the ability of the water to conduct electricity.

T F **8.** Boiler water tests and water treatment activities are recorded in a safety data sheet similar to a boiler room log.

Multiple Choice

______________ **1.** ___ is a measure of the acid-alkaline balance of a solution using a scale ranging from 0 to 14.

A. pH
B. Alkalinity
C. A hardness test
D. boiler water treatment

______________ **2.** A solution with a pH greater than ___ is alkaline.

A. 0
B. 7
C. 14
D. 21

______________ **3.** Boiler water pH requirements vary, but the pH value of boiler water typically should be maintained between ___ to prevent scale formation.

A. 5 and 7.5
B. 8 and 10.5
C. 10 and 11.5
D. 12 and 14.5

_______________ **4.** High alkalinity can lead to ___.
A. carryover
B. foaming
C. bottom blowdown
D. caustic embrittlement

_______________ **5.** A ___ test is a water treatment test using a reagent to determine the concentration of a specific dissolved substance.
A. soap
B. titration
C. conductivity
D. pH

Water Treatment Log Readings

_______________ **1.** Third day of month – condensate pH

_______________ **2.** Sixth day of month – sodium sulfite present

_______________ **3.** Second day of month – condensate TDS

_______________ **4.** Fifth day of month – condensate hardness

_______________ **5.** Fourth day of month – boiler water TDS

Heating and Chilling Plant
January
Boiler #1

Date	Boiler Water					Feedwater	Condensate			Products				Blowdown
	P	M	OH	TDS	Na_2SO_3	TDS	pH	TDS	Hard	938	8570	960	9980	
Max			400	3500	60		10.8							
Min			200	2500	30		10.0							
1	320	384	256	2700	45	45	10.8	15	0	44	4	4	12	3 / 1 hr
2	376	440	312	2900	60	39	10.8	16	0	32	8	0	0	3 / 2 hr
3	348	400	296	3000	55	36	10.9	16	0	32	6	0	8	3
4	324	380	268	2700	40	39	10.5	14	0	32	6	4	4	3
5	340	392	288	2900	45	34	10.7	14	0	32	6	4	8	3
6	272	328	216	2300	45	36	10.6	13	0	32	8	4	8	–
7	290	364	228	2600	35	34	10.8	13	0	36	4	4	8	–
8	311	396	336	2700	50	39	10.5	13	0	40	8	0	4	3

Chapter 7

Boiler Water Treatment

Section 7.3—Boiler Water Treatment

Name ______________________ **Date** ______________

True-False

T F 1. Minimizing the makeup water added to a boiler reduces the water treatment chemicals required.

T F 2. Internal boiler water treatment involves treating water before it enters a boiler.

T F 3. A deaerator is a boiler water pre-treatment accessory that is used to remove calcium and magnesium to reduce hardness.

T F 4. A package feedwater system can be single- or multiple-tank and automatically or manually operated and is used in conjunction with other external boiler water treatment or internal boiler water treatment systems.

T F 5. A deaerator is a boiler water treatment accessory that heats feedwater to the temperature of saturated steam at the pressure within a pressure vessel.

T F 6. Surge tanks are vented to the atmosphere and commonly lined with an epoxy coating to prevent corrosion.

T F 7. Bypass feeders are commonly used on larger boilers in plants where the load varies and makeup water demand is high.

T F 8. Equipment in chemical feed systems varies but usually includes a mixing tank, a pump, control valves, piping, and a control mechanism.

T F 9. Corrosion can be prevented by preheating the feedwater fed to a boiler to help remove oxygen and carbon dioxide.

T F 10. When continuous blowdown is used, bottom blowdown primarily removes suspended solids or sludge.

T F 11. A bottom blowdown is most effective when the boiler is under high load and the feedwater input is high.

T F 12. Improper water treatment is a common cause of boiler failure.

Multiple Choice

______________ 1. Pre-treatment equipment commonly includes ___.
- A. draft fans and surge tanks
- B. blowdown valves and vacuum pumps
- C. filters and dampers
- D. filters and water softeners

_______________ 2. A(n) ___ water softener softens water by replacing calcium and magnesium with sodium.
A. oxygen
B. sodium sulfate
C. zeolite
D. silica

_______________ 3. If plant conditions do not warrant a(n) ___, then a package feedwater system is commonly used.
A. deaerator
B. transformer
C. accumulator
D. circulating pump

_______________ 4. A ___ is used in a boiler system if there are intermittent peak loads of condensate that can exceed the surge capacity of the deaerator, such as condensate with varying pressures or temperatures or condensate that does not have sufficient pressure to enter the deaerator on its own.
A. pressure regulating valve
B. surge tank
C. bypass feeder
D. compression tank

_______________ 5. A(n) ___ is a water treatment accessory that introduces water treatment chemicals in a loop.
A. bypass feeder
B. os&y valve
C. circulating pump
D. surge tank

_______________ 6. Feedwater chemicals are mixed in the ___.
A. return tank
B. surge tank
C. deaerator
D. mixing tank

_______________ 7. ___ is the process of evacuating (discharging) water and undesirable accumulated material from a boiler.
A. Carryover
B. Blowdown
C. Priming
D. Atomizing

_______________ 8. A quick-open bottom blowdown valve is located ___.
A. on the continuous blowdown line
B. between the two slow-opening valves
C. closest to the boiler
D. farthest from the boiler

Bypass Feeders

_______________ **1.** Bypass feeder tank

_______________ **2.** Feedwater pump

_______________ **3.** Stop valve

_______________ **4.** Water treatment chemicals added

_______________ **5.** Check valve

_______________ **6.** Boiler

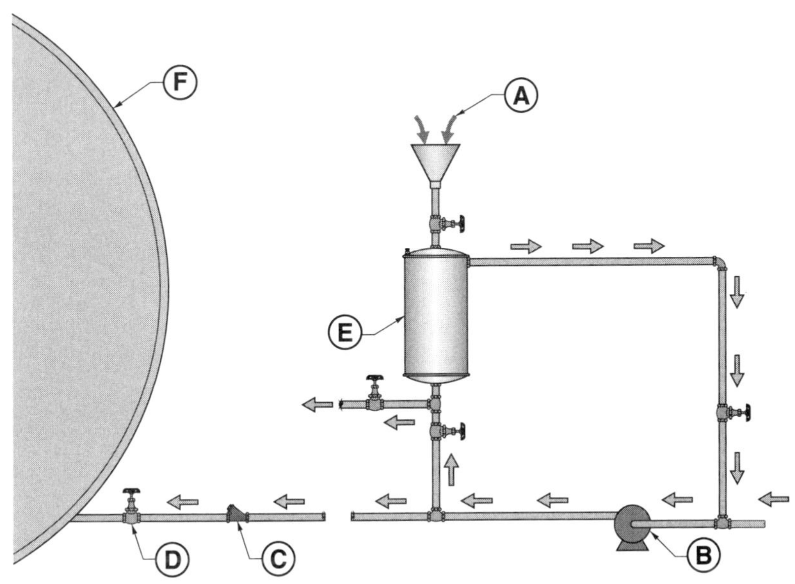

 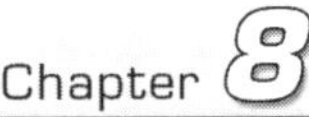

Boiler Operation Procedures

Section 8.1—Taking Over a Shift

Name ______________________________ **Date** ______________________

True-False

T F **1.** When taking over a shift, preliminary safety checks allow the boiler operator to identify any possible problems.

T F **2.** A shift log helps a boiler operator ensure that critical duties, such as safety valve tests, are performed.

T F **3.** In general, steam boilers should be blown down every 24 hours.

T F **4.** The water level in a gauge glass should be out of sight during bottom blowdown.

T F **5.** A boiler should be blown down when it is on a heavy load.

T F **6.** During bottom blowdown, the boiler operator should hold the valve handle until the valve closes.

T F **7.** A blowdown valve should be kept open long enough for the water level in the gauge glass to drop out of sight.

T F **8.** The quick-opening valve is opened first and closed last during blowdown.

T F **9.** An evaporation test is performed by blowing down the low water fuel cutoff.

T F **10.** A flame scanner is tested with a burner firing to simulate a flame failure.

T F **11.** After blowing down the water column and gauge glass, the low water fuel cutoff should be blown down also.

T F **12.** Procedures commonly performed and logged during a shift include water column and gauge glass blowdown, bottom blowdown, low water fuel cutoff testing, and flame scanner testing.

Multiple Choice

______________ **1.** The first thing a boiler operator should do when entering a boiler room is ___.
- A. read the boiler room log
- B. check the fuel supply
- C. blow down the flash tank
- D. check the boiler water level

______________ **2.** The boiler should be blown down when it is ___.
- A. in high fire
- B. shut down
- C. on a light load
- D. starting up

__________ **3.** The ___ valve should be closest to the boiler.
A. screw
B. quick-opening
C. os&y
D. globe

__________ **4.** After blowing down the water column and gauge glass, the water should ___ the gauge glass.
A. flash to steam as it enters
B. quickly enter
C. slowly enter
D. not enter

__________ **5.** A(n) ___ test is used to test the low water fuel cutoff.
A. hydrostatic
B. evaporation
C. dry layup
D. vacuum

__________ **6.** The boiler must be ___ when the low water fuel cutoff is blown down.
A. shut down
B. dumped
C. firing
D. in battery

__________ **7.** A(n) ___ test is the most accurate method of testing a low water fuel cutoff.
A. surface blowdown
B. hydrostatic
C. vacuum
D. evaporation

__________ **8.** When testing a low water fuel cutoff, the burner should shut off when ___.
A. water is still visible in the gauge glass
B. the gauge glass is empty
C. the float is removed
D. water is no longer visible in the gauge glass

__________ **9.** A boiler is given a bottom blowdown to ___.
A. remove sludge
B. control low water
C. increase chemical concentrations
D. reduce floating particulate

__________ **10.** A low water fuel cutoff should be tested ___.
A. daily
B. weekly
C. monthly
D. annually

Chapter 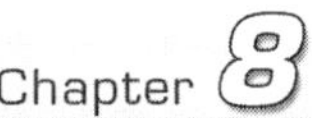

Boiler Operation Procedures

Section 8.2—Startup and Shutdown

Name ______________________ **Date** ______________

True-False

T F **1.** A main steam stop valve should be closed during normal plant start-up.

T F **2.** A boiler vent must be closed during warm-up to prevent pressure from escaping.

T F **3.** A brick refractory should be cooled rapidly for energy efficiency.

T F **4.** A boiler should be filled with water heated to 70°F before it is started for the first time.

T F **5.** A hydrostatic test is a steam test used to check for boiler leaks.

T F **6.** During a hydrostatic test, the boiler is half filled with water.

T F **7.** When warming up a boiler, the low water fuel cutoff should be blown down before the steam pressure gauge indicates pressure in the boiler.

T F **8.** When shutting down a boiler, the main steam valve should be closed immediately after the furnace is shut down.

Multiple Choice

______________ **1.** The boiler pressure should always be ___ the header pressure when cutting a boiler in on the line.

A. slightly above
B. slightly below
C. the same as
D. 50 psi below

______________ **2.** A(n) ___ test is used to check for leaks in a boiler.

A. evaporation
B. low water
C. caustic
D. hydrostatic

______________ **3.** To prevent damage to a brick refractory, the furnace ___.

A. should be cooled slowly
B. is lit off in high fire
C. is purged before shutdown
D. should be purged after shutdown

____________________ **4.** The water column whistle valve is ___ during a hydrostatic test.

A. removed

B. activated once

C. sounded

D. replaced with a new valve

____________________ **5.** During a hydrostatic test, pressure on the boiler is brought up to ___ times the MAWP.

A. 1½

B. 2

C. 2½

D. 5

____________________ **6.** Water used to fill the boiler during a hydrostatic test should be a minimum of ___°F.

A. 0

B. 12

C. 32

D. 70

Chapter 8

Boiler Operation Procedures

Section 8.3—Maintenance

Name ______________________ Date ______________

True-False

T F 1. A boiler should be dumped when hot to allow the heat to be recovered.

T F 2. A boiler must be taken off-line before it can be inspected.

T F 3. Main steam stop valves must be locked out and tagged during a boiler inspection.

T F 4. After a boiler is dumped, it should dry completely before cleaning.

T F 5. Checking the gauge glass is the quickest way to determine water level in a boiler.

T F 6. A float ball in a low water fuel cutoff should never be polished.

T F 7. Dry layup is recommended for boilers that will be out of service for an extended period.

T F 8. In wet layup, the water side must be dry to prevent corrosion and pitting from residual moisture.

T F 9. More pounds of silica gel than quicklime are required for the dry layup of a boiler.

T F 10. All valves are opened during dry layup to prevent moisture from entering the boiler.

T F 11. A scratch on the inside of a gauge glass can cause it to break.

T F 12. A clean gauge glass is the only way to accurately determine boiler water level.

T F 13. The plant manager is responsible for performing boiler maintenance tasks using established procedures.

T F 14. All plants maintain boiler room logs for an 8-hour period.

T F 15. A boiler room log can be used to determine the cause of a boiler shutdown.

T F 16. Boiler layup is required when a boiler is out of service between shifts.

T F 17. Any maintenance and cleaning of the low water fuel cutoff must be done while there is no pressure on the boiler.

T F 18. Improper feedwater treatment is probably the largest direct cause of boiler failure.

Multiple Choice

______________ 1. When water and sulfur combine, they react to form ___, which corrodes boiler metal.

A. sodium sulfite
B. zeolite
C. sulfuric acid
D. silica gel

_______________ **2.** A ___ is used to record information regarding operation of a boiler during a given period of time.

A. chief instruction handbook
B. pressure control
C. pressure gauge
D. boiler room log

_______________ **3.** Before dumping a boiler, the boiler heating surface should be ___.

A. in high fire
B. cool enough to touch
C. higher than 337°F
D. in low fire

_______________ **4.** ___ can be placed on desiccant trays inside the drums or inside the shell to absorb moisture from the air during dry layup of a boiler.

A. Anthracite
B. Quicklime
C. Sulfite
D. Sulfate

_______________ **5.** The internal mechanism of a low water fuel cutoff should be removed from the bowl at the recommended intervals to check and ___ the float ball.

A. clean
B. polish
C. scale
D. drain

_______________ **6.** New ___ should be used when replacing a broken gauge glass.

A. packing nuts
B. glands
C. seats
D. washers

_______________ **7.** During a boiler inspection, the main steam stop valve is ___.

A. opened and tagged out
B. closed, locked out, and tagged out
C. opened partially to allow steam flow
D. fully open to allow inspection of the valve

_______________ **8.** Low water fuel cutoffs must be ___ for proper performance.

A. installed above the NOWL
B. protected by a siphon
C. vertically aligned
D. designed to withstand 10 times the MAWP

Boiler Operation Procedures

Section 8.4—Handling Boiler Conditions

Name ______________________________ **Date** ______________

True-False

T F **1.** Water added to a boiler with a low water level condition could cause a boiler explosion.

T F **2.** According to incident reports from The National Board of Boiler and Pressure Vessel Inspectors, the most frequent causes of boiler incidents are operator error and poor maintenance.

T F **3.** Even though it is considered a low water condition to have water below the NOWL, it does not become dangerous until the water in the gauge glass drops out of sight.

T F **4.** A low water condition is caused by an excessive amount of water in a boiler as indicated by the gauge glass.

T F **5.** If there is an overpressure condition in a boiler, then the water level controls have failed.

T F **6.** Flame failure can cause a furnace explosion from the buildup and ignition of fuel.

T F **7.** To correct a steambound feedwater pump, water fed to the pump must be heated.

T F **8.** Before relighting a burner, the combustion control programmer must go through a purge cycle.

Multiple Choice

______________ **1.** A furnace explosion can be caused by ___.

A. a failed low water fuel cutoff
B. a plugged safety valve
C. a failed steam pressure reducing valve
D. leaking fuel

______________ **2.** If accessible, gas lines should be checked for leaks by ___.

A. purging the furnace
B. lighting a match
C. applying a mixture of soapy water
D. replacing the butterfly valve

______________ **3.** A steambound pump is caused by ___.

A. excessive water temperature
B. improper feedwater treatment
C. low furnace temperature
D. low water temperature

_______________ **4.** Boiler ___ is a condition that occurs when a boiler is operating at or above its maximum allowable working pressure.

A. NOWL
B. MAWP
C. overpressure
D. superheating

_______________ **5.** To correct a high water condition, a ___ is performed to bring the water level to the NOWL.

A. low water fuel cutoff test
B. surface blowdown
C. bottom blowdown
D. boiler vent test

_______________ **6.** Flame failures in gas-fired boilers are commonly caused by ___.

A. leaking fuel oil valves
B. insufficient pressure in gas lines
C. low water temperature
D. damaged feedwater pumps

_______________ **7.** Furnace explosions are normally investigated by ___.

A. a plant manager
B. a boiler operator
C. a fire inspector
D. the boiler manufacturer

Chapter 

Hot Water Boilers and Fittings

Section 9.1—Hot Water Heating Systems

Name ______________________________ **Date** ______________

True-False

T F **1.** Hot water heating systems produce heat more consistently than steam heating systems.

T F **2.** Hot water heating systems require less water than steam to transfer the same amount of heat.

T F **3.** A natural-circulation hot water heating system uses circulating pumps to transport water in the system.

T F **4.** Water becomes less dense as it is heated.

T F **5.** A natural-circulation hot water heating system uses a compression tank to absorb pressure changes.

T F **6.** Natural-circulation hot water heating systems are commonly used in large contemporary buildings.

T F **7.** A forced-circulation hot water heating system is vented to the atmosphere.

T F **8.** Water contains less heat energy (by weight) than steam.

T F **9.** Hot water heating systems transfer heat by circulating heated water to and from a building space.

T F **10.** An expansion tank moves water from a boiler to a supply line.

Multiple Choice

______________ **1.** A typical steam heating system operates at temperatures above ___°F.

A. 150
B. 170
C. 212
D. 300

______________ **2.** A hot water heating system operates at ___.

A. a higher temperature than a steam heating system
B. a lower temperature than a steam heating system
C. the same temperature as a steam heating system
D. a constant temperature all year

_____________________ **3.** In a natural-circulation hot water heating system, the ___ functions as a relief valve.
A. supply line
B. return line
C. heating unit
D. expansion tank

_____________________ **4.** A hot water heating system is isolated from the makeup water supply line by a ___.
A. backflow preventer
B. return line
C. circulating pump
D. low water fuel cutoff

Natural-Circulation Hot Water Heating Systems

_____________________ **1.** Vent line

_____________________ **2.** Heating unit

_____________________ **3.** Overflow line

_____________________ **4.** Boiler

_____________________ **5.** Expansion tank

_____________________ **6.** Branch line

_____________________ **7.** Return line

_____________________ **8.** Supply line

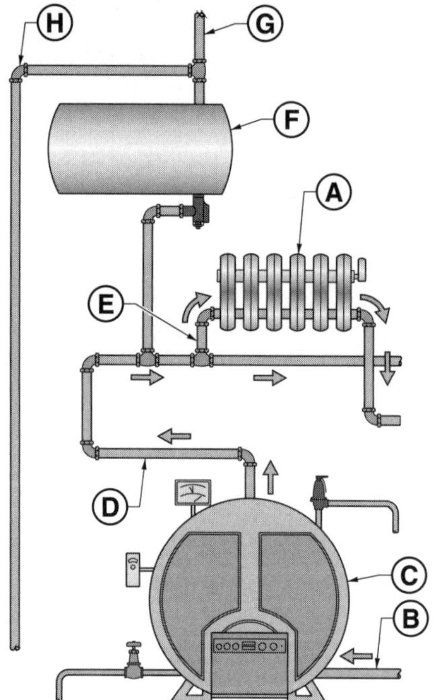

Forced-Circulation Hot Water Heating Systems

______ **1.** Air separator

______ **2.** Heating unit

______ **3.** Temperature-pressure gauge

______ **4.** Drain

______ **5.** Compression tank

______ **6.** Backflow preventer

______ **7.** Boiler

______ **8.** Compression tank valve

______ **9.** Air vent

______ **10.** Aquastat

______ **11.** Bypass valve

______ **12.** Stop valve

______ **13.** Air control tank fitting

______ **14.** Flow control valve

______ **15.** Makeup water supply line

______ **16.** Pressure-reducing valve

______ **17.** Diverter fitting

______ **18.** Relief valve

______ **19.** Bypass line

______ **20.** Circulating pump

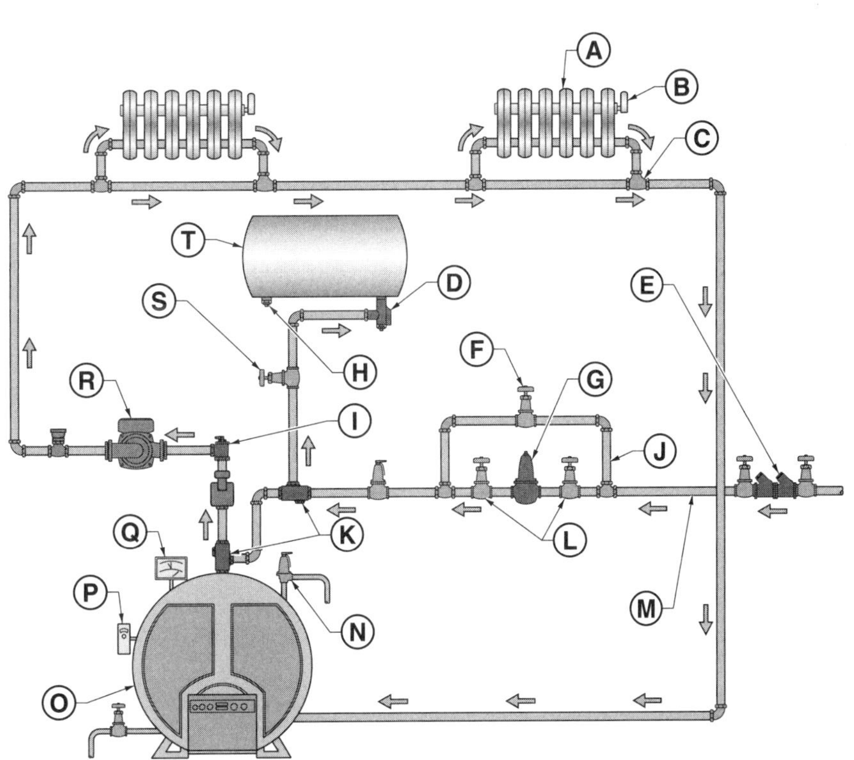

Chapter 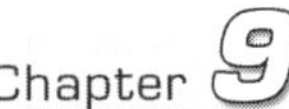

Hot Water Boilers and Fittings

Section 9.2—Hot Water Boilers

Name ______________________________ **Date** ______________

True-False

T F **1.** All boilers are manufactured in conformance with Section I or IV of the ASME Boiler and Pressure Vessel Code.

T F **2.** A condensing boiler achieves high efficiency by condensing the vapor in exhaust gases and recovering its latent heat of vaporization.

T F **3.** Steady-state efficiency is a measure of how efficiently a boiler's burner burns fuel without considering any losses.

T F **4.** The determining factor whether a boiler is in condensing mode or not is the return water temperature.

T F **5.** A boiler has a lower efficiency when it operates longer without shutdown.

T F **6.** Most noncondensing boilers could be forced to condense by changing the set points of the controls.

T F **7.** Electric boilers are classified by water volume and wattage.

T F **8.** A tankless boiler has an added heat exchanger to remove the required additional heat from the gases of combustion in order to allow the water vapor to condense.

Multiple Choice

______________ **1.** Steam boilers are classified as high pressure if steam pressure exceeds ___ psi.
A. 7
B. 15
C. 20
D. 160

______________ **2.** Hot water boilers operating with a 250°F water temperature and ___ psi water pressure or less are classified as low pressure.
A. 10
B. 15
C. 100
D. 160

________________ **3.** In actual practice, most condensing boiler efficiencies are around ___%.
A. 50
B. 75
C. 90
D. 100

________________ **4.** ___ efficiency is a measure of how efficiently a boiler's burner burns fuel while accounting for radiation and convection losses.
A. Thermal
B. Combustion
C. Steady-state
D. Seasonal

________________ **5.** ___ boilers do not retain any water internally except for what is in the heat exchanger coil.
A. Electric
B. Tankless
C. Low pressure
D. High pressure

Hot Water Boilers and Fittings

Section 9.3—Burners and Igniters

Name ______________________________ Date ______________

True-False

T F 1. A burner is the heat-producing component of a combustion boiler where fuel is burned.

T F 2. A back-pressure burner is a burner consisting of tubes with drilled holes that allow pressurized gas to escape into the furnace where the gas is burned.

T F 3. An igniter is a device used to set fire to the fuel in a burner.

T F 4. Electric spark igniters are controlled by combustion safety control systems.

T F 5. An electric spark igniter is a device located at a burner that uses a small piece of silicon nitride, which produces high temperatures when electric current passes through it.

Multiple Choice

______________ 1. The two types of electric igniters are the ___ igniter and the hot surface igniter.
- A. power
- B. pilot
- C. electric spark
- D. tubular

______________ 2. A(n) ___ burner is a small burner located near burner tubes that ignites the air-fuel mixture released into the burner.
- A. pilot
- B. tubular
- C. power
- D. electric spark

______________ 3. A power burner is a burner ___.
- A. that uses heating elements to heat water
- B. that uses sparks to ignite fuel in the burner
- C. consisting of tubes with drilled holes that allow pressurized gas to escape
- D. with a forced air fan that brings air under pressure into close contact with the fuel

Chapter 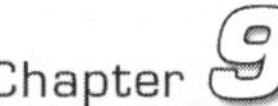

Hot Water Boilers and Fittings

Section 9.4—Hot Water Boiler Fittings

Name ______________________________ **Date** ____________________

True-False

T F **1.** An aquastat is a hot water boiler fitting that measures the water temperature in the boiler and controls the temperature by starting and stopping the burner.

T F **2.** An aquastat controls the burner by sensing steam pressure in a hot water boiler.

T F **3.** The relief valve should be located on the highest part of the boiler.

T F **4.** Relief valves should be manually tested every 30 days as recommended by the ASME Code.

T F **5.** A safety valve must pop fully open when the set pressure is reached and is used for gases and vapors only, but a safety relief valve can be used for gases, vapors, or liquids.

T F **6.** A temperature-pressure gauge has two scales.

T F **7.** An aquastat commonly has a temperature differential preset by the manufacturer.

T F **8.** The primary function of a pressure gauge is to de-energize a burner safety circuit and shut down the burner if the water level in the system drops below a safe operating level.

T F **9.** Low-water cutoffs must be checked at least weekly or when local ordinances require.

T F **10.** A water flow-proving switch is activated by the flow of water through the discharge piping from a circulating pump.

Multiple Choice

______________ **1.** An outside air thermostat measures the temperature of outside air, which is used to ___.

A. control the water temperature in a boiler
B. optimize draft airflow
C. balance the load on multiple boilers
D. adjust fuel flow

______________ **2.** A hot water boiler relief valve is rated in ___ relieved per hour.

A. British thermal units
B. pounds of water pressure
C. pounds of water temperature
D. pounds of flash steam

_______________ **3.** A relief valve must be built according to the ___ and must have specifications identified on a data plate.
A. ASME Code
B. ABMA Code
C. boiler manufacturer specifications
D. NFPA Code

_______________ **4.** The relieving pressure of a relief valve on a low pressure hot water boiler cannot exceed ___ psi or the MAWP designated on the boiler.
A. 15
B. 30
C. 160
D. 250

_______________ **5.** The two types of low water fuel cutoffs normally used on hot water boilers are the ___-type and the probe-type cutoffs.
A. electrostatic
B. electronic
C. float
D. flow

_______________ **6.** To protect pressure gauges from the damage caused by pressure surges, ___ can be used.
A. plugs
B. corks
C. snubbers
D. stoppers

_______________ **7.** A temperature-pressure gauge indicates the ___.
A. temperature and pressure of water leaving the boiler
B. temperature and pressure of the gases of combustion
C. air temperature and the draft pressure
D. draft temperature and the air pressure

_______________ **8.** Altitude pressure gauges typically have one black needle and one ___ needle.
A. blue
B. yellow
C. green
D. red

_______________ **9.** A(n) ___ is a hot water boiler fitting that serves as a boiler water setpoint temperature control system with an outside air thermostat.
A. relief valve
B. hot water reset control
C. pressure gauge
D. aquastat

______________________ **10.** A(n) ___ is not intended for hot water service.
A. aquastat
B. safety valve
C. safety relief valve
D. low water fuel cutoff

Hot Water Reset Controls

______________________ **1.** Air separator

______________________ **2.** Setpoint adjustor

______________________ **3.** Thermostat

______________________ **4.** Boiler

______________________ **5.** Cover

______________________ **6.** Remote bulb sensor

______________________ **7.** Aquastat

______________________ **8.** Burner controls

______________________ **9.** Heating units

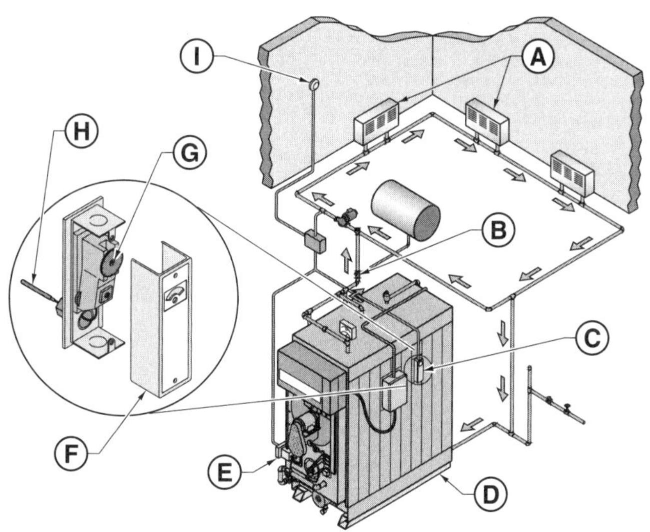

Relief Valves

_______________ **1.** Sealing diaphragm

_______________ **2.** Body

_______________ **3.** Outlet

_______________ **4.** Spring

_______________ **5.** Data plate

_______________ **6.** Try lever

_______________ **7.** Inlet

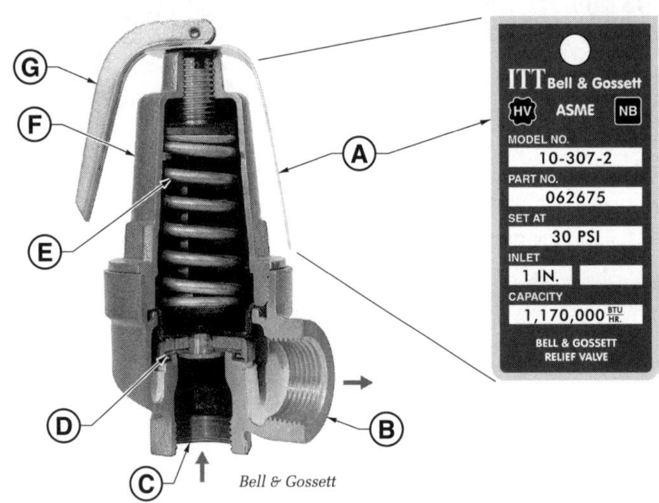

Chapter 10

Hot Water Boiler Accessories and Piping Systems

Section 10.1—Hot Water Boiler Accessories

Name ______________________ Date ______________

True-False

T F 1. Floor-mounted circulating pumps are smaller pumps that do not have piping running to and from them.

T F 2. Pipe-mounted circulating pumps are smaller pumps that are attached directly to the piping.

T F 3. A pressure-reducing valve is a hot water boiler accessory that reduces incoming makeup water pressure to the system operating pressure.

T F 4. Stop valves before and after a pressure-reducing valve allow servicing of the pressure-reducing valve.

T F 5. A strainer should be installed before an RPZ backflow preventer to remove any foreign matter that could cause the preventer to malfunction.

T F 6. A compression tank is normally full of water to feed into the boiler during a low water condition.

T F 7. The three designs of strainers used in hot water heating systems are basket, in-line, and wye.

T F 8. A backflow preventer with an intermediate atmospheric vent consists of four independent check valves within the body and an atmospheric vent between the check valves.

T F 9. A compression tank can be located either after the boiler (outlet side) or before the boiler (inlet side).

T F 10. A common cause of problems in a hot water system is air in the system.

Multiple Choice

______________ 1. A ___ tank is a hot water system tank used to absorb changes in water pressure in a forced-circulation hot water heating system.

A. blowdown
B. compression
C. receiver
D. vacuum

_______________ **2.** A pressure-reducing valve is set to reduce water pressures to ___ psi.
- A. 4
- B. 8
- C. 10
- D. 12

_______________ **3.** A ___ consists of two independent spring-loaded check valves, test cocks, and isolation valves.
- A. flow control valve
- B. water strainer
- C. reduced-pressure zone backflow preventer
- D. combination heating system

_______________ **4.** A ___ is a dynamic pump with a rotating impeller inside a cast iron or steel housing.
- A. pressure-reducing valve
- B. centrifugal pump
- C. balancing valve
- D. gear pump

_______________ **5.** Some compression tanks are equipped with a ___ for determining the compression tank water level.
- A. sight glass
- B. vent
- C. gauge glass
- D. relief valve

Hot Water Boiler Accessories and Piping Systems

Section 10.2—Hot Water System Valves

Name ______________________________ **Date** ____________________

True-False

T F **1.** When a circulating pump is not operating, the flow control valve closes to prevent natural circulation.

T F **2.** Diverter fittings are used to supply cold water to a heating unit while maintaining the required flow of cold water to the next heating unit.

T F **3.** Check valves are available as swing, lift, and spring check valves.

T F **4.** Common types of hot water system valves include diverter fittings, three-way valves, and centrifugal valves.

T F **5.** The temperature range of a thermostatic drain valve is normally adjustable between 43°F and 79°F.

Multiple Choice

______________ **1.** A flow control valve functions similarly to a(n) ___ valve to prevent natural water circulation.

A. air separator
B. pressure reducing
C. check
D. gate

______________ **2.** A ___ is a hot water boiler accessory resembling a piping tee that meters water flow from the supply line to the heating unit.

A. flow control valve
B. hand-operated valve
C. steam trap
D. diverter fitting

______________ **3.** A ___ check valve is a check valve that prevents backflow through the use of a conical or cylindrical brass disc held in place by a spring.

A. flow control
B. spring
C. lift
D. swing

_______________ **4.** Hot water boilers use ___ instead of the bottom blowdown valves used with steam boilers.
A. drain valves
B. diverter fittings
C. backflow preventers
D. compression tanks

_______________ **5.** A ___ valve is a hot water system control valve that regulates the rate of fluid flow through a hot water piping system to meet system design requirements.
A. three-way
B. swing check
C. balancing
D. flow control

Low Pressure Boilers

Chapter 

Hot Water Boiler Accessories and Piping Systems

Section 10.3—Piping Systems

Name ______________________________ **Date** ____________________

True-False

T F **1.** Piping system designs vary depending on the degree of temperature control desired and the specific application.

T F **2.** The two common types of two-pipe systems are series systems and primary-secondary systems.

T F **3.** One-pipe systems include direct-return and reverse-return systems.

T F **4.** Balancing the water flow in a two-pipe reverse-return system is much easier than in other types of systems because the resistance to the water flow is the same for all heating units.

T F **5.** A combination heating system uses all the components of a hot water heating system and a steam heating system.

Multiple Choice

______________ **1.** In a ___ system, the inlet temperature at each consecutive heating unit is lower than at the previous heating unit.

A. one-pipe primary-secondary
B. one-pipe series
C. two-pipe
D. four-pipe

______________ **2.** ___ systems are difficult to balance because the water that flows through the last heating unit must be pumped back through the whole piping system to the boiler.

A. Two-pipe direct-return
B. Two-pipe reverse-return
C. Three-pipe hot water heating and cooling
D. Combination heating

______________ **3.** ___ -pipe systems are used when different parts of a system require heating and cooling simultaneously.

A. One
B. Two
C. Three
D. Combination one- and two

______________ **4.** A ___ -pipe system completely separates heating and cooling supply and return lines.

A. one
B. two
C. three
D. four

______________ **5.** In a ___ system, a boiler is used to produce the steam required for building processes.

A. combination heating
B. one-pipe series
C. two-pipe reverse-return
D. four-pipe hot water heating and cooling

Cooling Systems

Section 11.1—Cooling System Principles

Name ______________________ **Date** ______________

True-False

T F **1.** Water transports heat more efficiently with less temperature loss over distances than air.

T F **2.** Some cooling systems can be designed to operate as a heating system or cooling system and share the same components.

T F **3.** Common cooling system applications include air conditioning of building spaces and refrigeration in food processing, such as in a pasteurization process.

T F **4.** Refrigeration is the process of moving heat from an area where it is undesirable to an area where the heat is acceptable.

T F **5.** A measurement of the quantity of heat transfer is the British thermal unit (Btu).

T F **6.** Latent heat is heat identified by a change of state and temperature change of a substance.

T F **7.** When heat is added to a liquid, the liquid hardens and becomes a solid.

T F **8.** Specific heat has no unit of measurement since it is a ratio, but it is almost identical numerically to the heat capacity because of the way the Btu is defined.

T F **9.** Radiation has a major impact on all cooling systems.

T F **10.** In a refrigeration system, the specific volume of the refrigerant is used in determining the size of the required compressor.

T F **11.** As the pressure increases in a boiler, the boiling and condensing point of the water decrease.

T F **12.** The specific gravity of water is 2.0.

Multiple Choice

______________ **1.** ___ heat must be added or removed to cause a substance to change state between a solid and a liquid or a liquid and a gas.

A. Super
B. Sensible
C. Latent
D. Mechanical

__________ **2.** A(n) ___ is a cooling system that chills water for cooling.
- A. chiller
- B. refrigeration system
- C. ductwork system
- D. air conditioning system

__________ **3.** The three states of ___ are solid, liquid, and gas.
- A. gravity
- B. density
- C. volume
- D. matter

__________ **4.** ___ heat is the ratio of the heat capacity of a substance to the heat capacity of a standard substance at the same temperature.
- A. Latent
- B. Specific
- C. Mechanical
- D. Super

__________ **5.** ___ heat is heat that causes a temperature change that can be measured with a thermometer or sensed by a person.
- A. Sensible
- B. Super
- C. Latent
- D. Specific

Cooling System Application – Pasteurization Process

__________ **1.** Cooled milk

__________ **2.** Warm water out

__________ **3.** Collection trough

__________ **4.** Heated milk enters cooler

__________ **5.** Milk passing around piping

__________ **6.** Chilled water in

__________ **7.** Piping containing chilled water

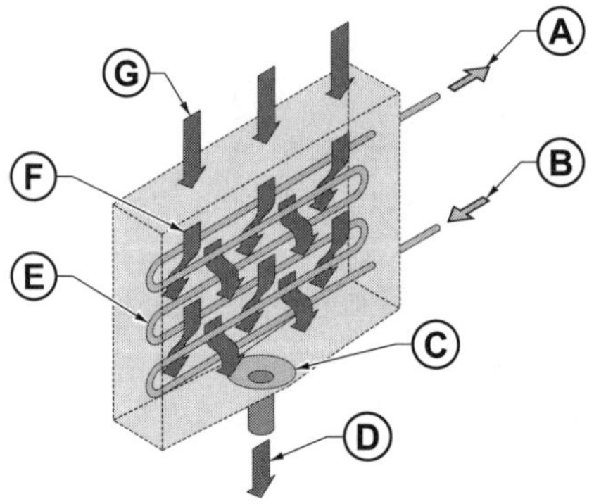

Chapter 

Cooling Systems

Section 11.2—Cooling Systems

Name ______________________________ **Date** ______________

True-False

T F **1.** The two common types of cooling systems are direct-expansion systems and indirect expansion systems.

T F **2.** The main difference between how cooling systems operate compared to hot water heating systems is that cooling systems transport heat away from a building space to produce a cooling effect.

T F **3.** Cooling systems follow a cycle where the refrigerant moves through the five main system components.

T F **4.** A reciprocating compressor is a compressor that has reciprocating cylinders to compress refrigerant.

T F **5.** Centrifugal compressors generally do not have a great deal of compressive power, but they can handle a large volume of refrigerant.

T F **6.** In a direct-expansion system, the condenser can be a shell-and-coil, a shell-and-tube, or a tube-in-tube heat exchanger.

T F **7.** Flash gas is a refrigerant that boils off and cools liquid refrigerant when some of the refrigerant molecules absorb heat from the other refrigerant molecules.

T F **8.** A TXV maintains a constant superheat of about 8°F to 12°F.

T F **9.** Superheated vapor is vapor at a higher temperature than its corresponding pressure.

T F **10.** The four types of evaporators normally used in refrigeration systems are the air-to-refrigerant coil, shell-and-tube heat exchanger, tube-in-a-tube, and shell-and-coil.

T F **11.** The liquid line is a refrigerant line that carries the warm, high-pressure liquid refrigerant between a condenser and a metering device.

T F **12.** Additional components that may be encountered with cooling systems are liquid receivers, accumulators, and purge units.

T F **13.** A suction accumulator is a device used to remove air and moisture from a refrigeration system.

T F **14.** The two classifications of air conditioners are package air conditioners and split-system air conditioners.

T F **15.** Chilled water systems are classified as indirect cooling systems because chilled water is used as a medium to transfer heat and cool a building space.

T F **16.** Ammonia-water absorption chiller systems are commonly used for large applications where mechanical compression systems are not as efficient.

T F **17.** An ammonia-water chiller system commonly includes a generator, condenser, evaporator, and absorber.

T F **18.** A lithium bromide-water absorption chiller system is an absorption chiller system that uses water as the refrigerant and a solution of lithium bromide and water as the absorbent.

Multiple Choice

______________ **1.** A ___ is a rotary compressor that has rotating impellers to pressurize refrigerant.
A. boiler
B. chiller
C. condenser
D. centrifugal compressor

______________ **2.** High-pressure chillers have an operating pressure greater than ___ psi.
A. 10
B. 15
C. 25
D. 30

______________ **3.** A(n) ___ is a heat exchanger that removes heat from a high-pressure refrigerant vapor.
A. condenser
B. pump
C. expansion valve
D. boiler

______________ **4.** A ___ is a component of a refrigerant system that controls the flow of refrigerant into an evaporator.
A. generator
B. compressor
C. metering device
D. chiller

______________ **5.** A(n) ___ is a metering device that senses the temperature of the refrigerant discharged from an evaporator and controls the amount of liquid refrigerant flowing into the evaporator.
A. thermostatic expansion valve
B. electronic expansion valve
C. high-side float
D. low-side float

_______________ **6.** A(n) ___ metering device is a metering device with a fixed opening that acts as a restriction in the liquid line between a condenser and an evaporator.
A. high-side float
B. low-side float
C. orifice
D. electronic expansion valve

_______________ **7.** A(n) ___ is a heat exchanger used to absorb heat from an area to be cooled and transfer the heat into refrigerant.
A. evaporator
B. compressor
C. generator
D. condenser

_______________ **8.** The ___ line is a refrigerant line that carries low-pressure refrigerant between an evaporator and a compressor.
A. decompression
B. liquid
C. hot gas discharge
D. suction

_______________ **9.** A(n) ___ is a container located between an evaporator and a compressor that catches any liquid refrigerant before it reaches the compressor and causes damage.
A. liquid receiver
B. evaporation tank
C. suction accumulator
D. condenser

_______________ **10.** A(n) ___ is a cooling system that uses direct expansion of the refrigerant in the evaporator coil to cool a building's interior air.
A. air conditioner
B. cooling tower
C. generator
D. air vent

_______________ **11.** ___ systems are commonly used to cool large buildings such as hotels and stadiums.
A. Absorption chiller
B. Chilled water
C. Bromide-water absorption chiller
D. Ammonia-water absorption chiller

_______________ **12.** Water in a chilled water system is typically cooled to about ___°F in the chiller.
A. 15
B. 30
C. 45
D. 60

_______________ **13.** Water cannot be used as a secondary refrigerant in any application where the temperature is below ___°F.
A. 32
B. 52
C. 72
D. 92

_______________ **14.** A(n) ___ is an absorption chiller system component that vaporizes the refrigerant and separates it from the absorbent.
A. compressor
B. absorber
C. condenser
D. generator

Refrigeration Cycle

_______________ **1.** Cool air

_______________ **2.** Hot air

_______________ **3.** Metering device

_______________ **4.** Evaporator

_______________ **5.** Compressor

_______________ **6.** Condenser

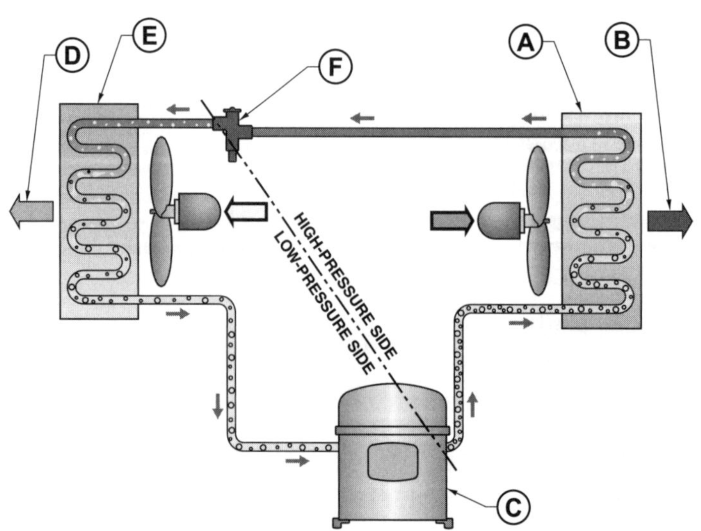

Centrifugal Compressors

____________________ **1.** Drive shaft

____________________ **2.** Housing

____________________ **3.** Impeller wheel

____________________ **4.** Inlet port

____________________ **5.** Outlet port

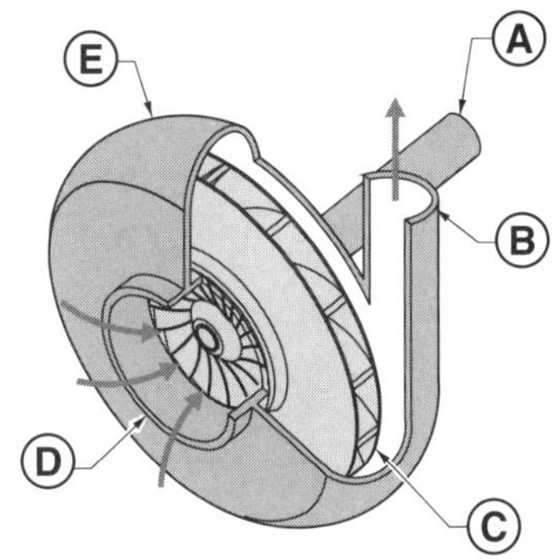

Ammonia-Water Absorption Chiller Systems

____________________ **1.** Generator

____________________ **2.** Expansion valve

____________________ **3.** Evaporator

____________________ **4.** Regulating valve

____________________ **5.** Absorber

____________________ **6.** Condenser

____________________ **7.** Pump

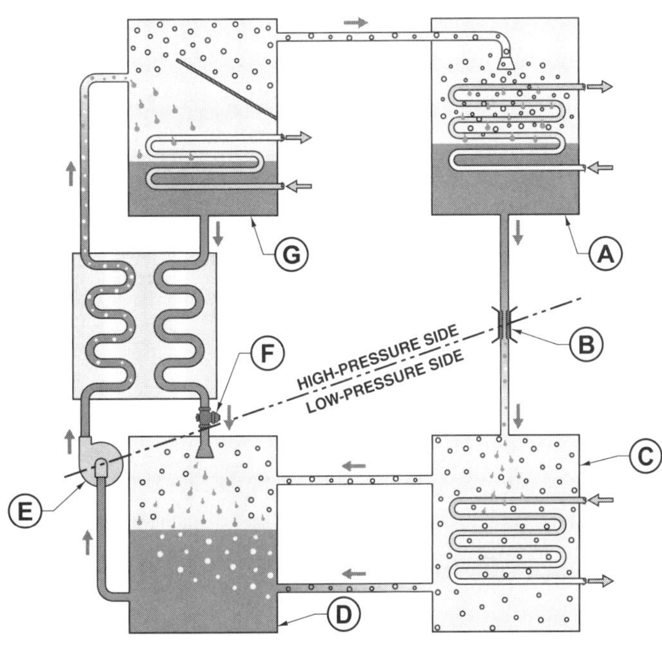

Chilled Water Systems

_______________ **1.** Compressor

_______________ **2.** Metering device

_______________ **3.** Cooling tower

_______________ **4.** Air duct

_______________ **5.** Evaporator

_______________ **6.** Ceiling

_______________ **7.** Condenser

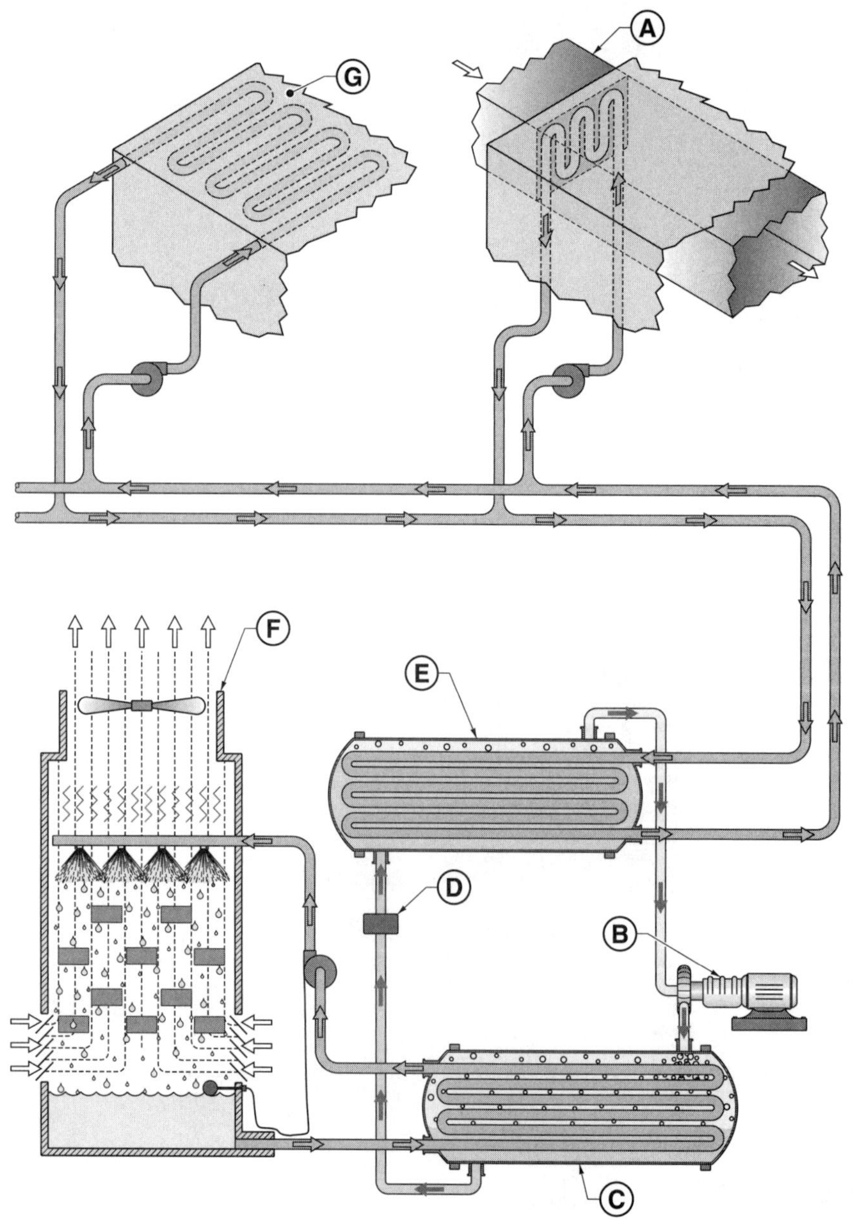

Chapter 11

Cooling Systems

Section 11.3—Refrigerants

Name ______________________________ **Date** ______________

True-False

T F **1.** Refrigerants are selected for their ability to absorb, transport, and release heat at the proper temperature and pressure.

T F **2.** Refrigerants used in mechanical refrigeration systems are derived from fluorine and bromine.

T F **3.** The three classifications of refrigerants most technicians work with today are chlorofluorocarbons (CFCs), hydrochloroflurorocarbons (HCFCs), and hydrofluorocarbons (HFCs).

T F **4.** The only techniques used for leak testing are the standing pressure method and the standing vacuum method.

T F **5.** Reclamation is the cleaning of refrigerant through oil separation and single or multiple passes through devices, such as replaceable core filter-driers.

T F **6.** When released to the atmosphere, refrigerants can cause frostbite to unprotected skin and freezing of the eyes, and they can create environmental problems in the form of ozone depletion and global warming.

Multiple Choice

______________ **1.** The ___ is charged with establishing and enforcing the Clean Air Act.

A. Occupational Safety and Health Administration (OSHA)
B. Environmental Protection Agency (EPA)
C. National Fire Protection Association (NFPA)
D. American National Standards Institute (ANSI)

______________ **2.** ___ certification is required for technicians servicing low-pressure appliances that use a refrigerant that boils above 50°F at atmospheric pressure.

A. Type I
B. Type II
C. Type III
D. Universal

_______________ **3.** Up to ___% annual leak rate is allowed for industrial and commercial refrigeration equipment before repair is required.

A. 15

B. 25

C. 35

D. 45

_______________ **4.** ___ is the removal of refrigerant in any condition from a system and the storage of the refrigerant in an external container without testing or processing it in any way.

A. Reuse

B. Reclamation

C. Recycling

D. Recovery

Low Pressure Boilers

Chapter 12 Boiler Operation Safety

Section 12.1—Regulatory Agencies and Standards Organizations

Name ______________________________ **Date** ____________________

True-False

T F **1.** Each year, all current OSHA standards are reproduced in the Code of Federal Regulations (CFR).

T F **2.** The National Fire Protection Association (NFPA) is a federal government agency established in 1970 to control and abate pollution in the areas of air, water, solid waste, pesticides, radiation, and toxic substances.

T F **3.** The Occupational Safety and Health Administration (OSHA) is an organization that promotes greater safety to life and property through uniformity in the construction, installation, repair, maintenance, and inspection of boilers and pressure vessels.

T F **4.** The purpose of the National Electrical Code® (NEC®) is the practical safeguarding of persons and property from hazards arising from the use of electricity.

T F **5.** The American National Standards Institute (ANSI) is a national standards organization that helps identify industrial and public needs for national standards.

T F **6.** U.S. Green Building Council® (USGBC) is an independent organization that tests equipment and products to verify conformance to national codes and standards.

T F **7.** Boiler manufacturers may be required to provide special boiler trim and accessories to meet insurance agency requirements.

T F **8.** The American Boiler Manufacturer's Association (ABMA) is a national organization that represents manufacturers of commercial, industrial, and utility steam generating and fuel burning equipment and suppliers.

Multiple Choice

______________ **1.** A ___ is an accepted reference or practice.

A. code
B. regulation
C. standard
D. requirement

______________ **2.** The ___ uses a four-color diamond-shaped sign to display basic information about hazardous materials.

A. RTK label
B. HMIG label
C. NFPA Hazard Signal System
D. safety data sheet

______________ **3.** The ___ is a nongovernmental international organization that is comprised of national standards institutions of over 90 countries.

A. EPA
B. ASME
C. NFPA
D. ISO

______________ **4.** The ___ promotes a whole-building approach to sustainability by recognizing performance in five key areas of human and environmental health.

A. USGBC
B. NFPA
C. ABMA
D. OSHA

Boiler Operation Safety

Section 12.2—Boiler Room Safety Programs

Name ______________________________ **Date** ______________________

True-False

T F **1.** A boiler room contains many combustible materials, and boiler room fire safety requires established procedures to reduce or eliminate conditions that could cause a fire.

T F **2.** Spontaneous combustion is combustion caused by self-ignition through chemical action.

T F **3.** Class D fires are fires that burn wood, paper, textiles, and other dry materials containing carbon.

T F **4.** A boiler is a confined space and requires special entry procedures.

T F **5.** A nonpermit-required confined space is a confined space that has specific health and safety hazards associated with it.

T F **6.** An entry permit must be posted at confined space entrances or otherwise made available to entrants before entering a permit-required confined space.

T F **7.** A hazardous material is a substance that could cause injury to personnel or damage to the environment.

T F **8.** A flammability hazard on the NFPA Hazard Signal System indicates the degree of susceptibility of materials to explode or release energy by themselves or by exposure to certain conditions or substances.

T F **9.** A safety data sheet (SDS) is an informational form used to relay chemical hazard information from the manufacturer, importer, or distributor to the employer.

T F **10.** A general boiler room safety rule says that when performing a bottom blowdown, the quick-opening valve should be opened first and closed last.

Multiple Choice

______________ **1.** Class ___ fires are fires that burn combustible metals such as zirconium, titanium, magnesium, sodium, and potassium.

A. A
B. B
C. C
D. D

________ **2.** A(n) ___ hazard area is a building or room that is used as a shop and related storage facility, light manufacturing plant, automobile showroom, or parking garage.
   A. extra
   B. ordinary
   C. light
   D. confined

________ **3.** The NFPA Hazard Signal System colors and numbers identify potential health (blue), flammability (red), reactivity (___), and special hazards (no special color).
   A. yellow
   B. orange
   C. green
   D. black

________ **4.** A ___ hazard on the NFPA Hazard Signal System indicates the degree of susceptibility of a material to burning based on the form or condition of the material and its surrounding environment.
   A. reactivity
   B. health
   C. flammability
   D. special

________ **5.** In ANSI Standard Z400.1/Z129.1-2006, Hazardous Workplace Chemicals—Hazard Evaluation and Safety Data Sheet Precautionary Labeling Preparation, Section ___, Hazard(s) Identification includes all hazards regarding the chemical and required label elements.
   A. 1
   B. 2
   C. 11
   D. 16

Hazardous Material Container Labeling – RTK Labeling

________ **1.** Health hazards

________ **2.** Signal word

________ **3.** Chemical or common name

________ **4.** Physical hazards

________ **5.** Handling and storage instructions

________ **6.** First aid procedures for exposure or contact

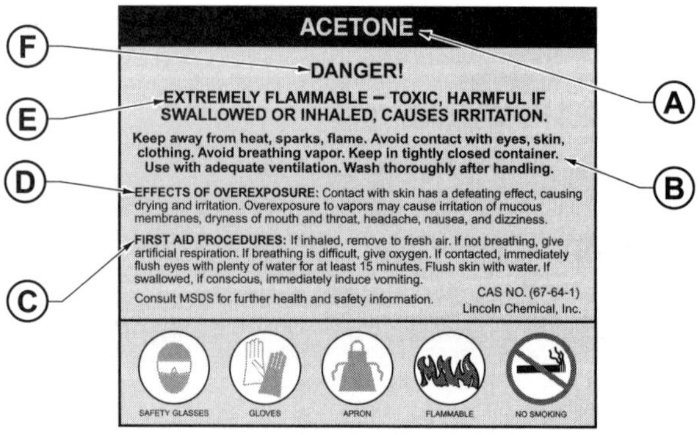

Boiler Operation Safety

Section 12.3—Personal Protective Equipment (PPE)

Name ______________________ **Date** ______________

True-False

T F **1.** Personal protective equipment (PPE) is a device or clothing worn by a boiler operator to prevent injury.

T F **2.** An ear muff is a device inserted into the ear canal to reduce the level of noise reaching the eardrum.

T F **3.** The work activities of the boiler operator determine the degree of dexterity required and the duration, frequency, and degree of exposure to potential hazards.

T F **4.** A face mask is a personal protective equipment device that provides respiratory protection from airborne contaminants.

Multiple Choice

______________ **1.** Protective ___ protect workers by resisting penetration and absorbing the blow of impact.
- A. harnesses
- B. gloves
- C. helmets
- D. earplugs

______________ **2.** Protective footwear must comply with___, *Personal Protection—Protective Footwear.*
- A. ANSI Z87.1-991
- B. ANSI Z41-1991
- C. OSHA 29 CFR 1910.133
- D. OSHA 29 CFR 1910.138

______________ **3.** A(n) ___ is a device worn over the ears to reduce the level of noise reaching the eardrum.
- A. protective shield
- B. helmet
- C. earplug
- D. ear muff

______________ **4.** ___ and waist belts are used to reduce injuries from falling.
- A. Full-body harnesses
- B. Safety ropes
- C. Protective barriers
- D. Personal protective equipment

Boiler Operation Safety

Section 12.4—Electrical Safety

Name ______________________________ **Date** ______________

True-False

T F **1.** An electrical safety policy must include the use of PPE.

T F **2.** The primary regulatory requirements related to electrical safety come from *NFPA 70E®, Standard for Electrical Safety in the Workplace.*

T F **3.** A safety inspector is a person who is trained in, and has specific knowledge of, the construction and operation of electrical equipment or a specific task, and is trained to recognize and avoid electrical hazards that might be present with respect to the equipment or specific task.

T F **4.** Proper maintenance of equipment is as critical to safety and proper operation of the equipment as the selection of PPE is to the protection of the worker.

T F **5.** Under all circumstances, electrical equipment should never be de-energized before maintenance work is performed.

T F **6.** After performing an electrical hazard analysis, tables in NFPA 70E are used to help determine the correct protective equipment for the expected exposure to the arc flash hazard.

Multiple Choice

______________ **1.** In the area of employer supervision, ___ states, "The employer shall determine, through regular supervision and through inspections conducted on at least an annual basis, that each employee is complying with the safety-related work practices required."

A. ANSI Z87.1-991
B. OSHA 29 CFR 1910.269
C. OSHA 29 CFR 1910.332
D. ANSI Z41-1991

______________ **2.** ___ states, "The training requirements contained in this section apply to employees who face a risk of electric shock that is not reduced to a safe level by the electrical installation."

A. ANSI Z41-1991
B. ANSI Z87.1-991
C. OSHA 29 CFR 1910.269
D. OSHA 29 CFR 1910.332

Low Pressure Boilers

Chapter 

Boiler Operation Safety

Section 12.5—General Safety Programs

Name ______________________ Date ______________

True-False

T F 1. Boiler room accidents can occur at any time.

T F 2. Accidents that occur must be reported regardless of the nature.

T F 3. Accident reports only include the name of the injured person and type of injury.

T F 4. General safety programs to reduce risk include lockouts and tagouts, emergency plans, and accident reports.

T F 5. A tagout prevents the startup of equipment and serves as a warning to operating and service personnel.

T F 6. A multiple lockout requires installation and removal of more than one lock on a multiple-lockout hasp.

T F 7. An authorized employee is responsible for developing and implementing an emergency plan.

T F 8. Boiler conditions that could lead to an emergency are high or low water conditions and flame failure.

Multiple Choice

______________ 1. ___ is the placement of a device on an energy-isolating device, in accordance with an established procedure, ensuring that the energy-isolating device and the equipment being controlled cannot be operated.

A. Safety locking
B. Shutdown tagging
C. Lockout
D. Tagout

______________ 2. A(n) ___ is a person who locks out or tags out machines or equipment in order to perform servicing or maintenance on that machine or equipment.

A. authorized employee
B. shift manager
C. boiler operator
D. designated plant official

_______________ **3.** A(n) ___ is a document that details procedures, exit routes, and assembly areas for facility personnel in the event of an emergency.

A. backup plan

B. escape route

C. accident report

D. emergency plan

Boiler Operator Licensing

Sample Exam 1

Name ______________________ Date ______________

True-False

T F 1. All steam boilers must have one or more safety valves.

T F 2. A hydrostatic test determines whether a steam boiler has sufficient relieving capacity.

T F 3. The main steam stop valve, bottom blowdown valves, and feedwater valves should be locked closed before anyone enters a boiler drum.

T F 4. Scale protects boiler heating surfaces.

T F 5. Boiler corrosion cannot be prevented as long as water is in contact with metal.

T F 6. Water begins to boil at approximately 212°F at atmospheric pressure.

T F 7. Fuels most commonly used in boilers are fuel oil and gas.

T F 8. Safety valves on steam boilers are designed to open slowly.

T F 9. A steam pressure gauge must be connected to the highest part of the steam side of a boiler.

T F 10. With an NOWL in a boiler, the gauge glass is approximately half full.

T F 11. A high surface tension on top of boiler water can lead to foaming.

Identify the ASME Code Symbol Stamps.

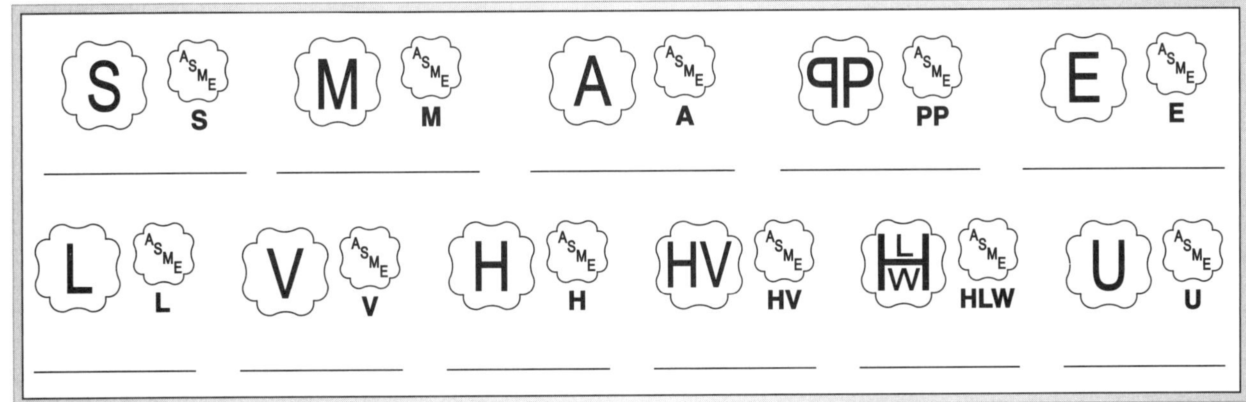

Multiple Choice

______________ **1.** The bottom blowdown of a boiler ___.
A. removes sludge and sediment from the mud drum
B. reduces boiler steam pressure
C. adds makeup water to the boiler
D. increases boiler priming

______________ **2.** A steam pressure gauge is calibrated in ___.
A. pounds per cubic inch
B. pounds per square inch
C. inches of water column
D. inches of water gauge

______________ **3.** Fires caused by spontaneous combustion are most likely to occur in the ___.
A. fuel oil tank
B. oily waste rag storage bin
C. boiler furnace
D. boiler ash pit

______________ **4.** A siphon installed between a boiler and a pressure gauge protects the Bourdon tube from ___ reaching the tube inside the gauge.
A. water
B. gases of combustion
C. steam
D. fuel

______________ **5.** Boiler feedwater is chemically treated to ___.
A. increase circulation
B. increase oxygen concentration
C. prevent formation of scale
D. increase boiler makeup water

______________ **6.** In a firetube boiler, soot accumulates on the ___.
A. inside tube surface
B. outside tube surface
C. waterwall surface
D. lowest part of the water side

______________ **7.** The pressure applied on the boiler during a hydrostatic test should be ___ times the MAWP.
A. 1½
B. 2
C. 2½
D. 3

______ **8.** As fuel oil is heated, its viscosity is ___.
A. the same
B. increased
C. decreased
D. ignited

______ **9.** An induced draft fan is located between a boiler and a ___.
A. feedwater pump
B. condensate return tank
C. evaporator
D. stack

______ **10.** The flash point of fuel oil is the minimum temperature at which the fuel will ___.
A. support combustion
B. no longer flow
C. flash when exposed to an open flame
D. have its highest Btu content

______ **11.** The fire point of fuel oil is the minimum temperature at which the fuel oil will ___.
A. burn continually
B. no longer flash
C. flash when exposed to an open flame
D. have its highest Btu content

______ **12.** The pour point of fuel oil is the lowest temperature at which fuel oil will ___.
A. burn
B. flow
C. ignite
D. pump

______ **13.** Fuel oil with a low flash point is ___.
A. only used with high pressure boilers
B. dangerous to handle
C. only used in low pressure plants
D. heated to decrease its viscosity

______ **14.** The operating range of a steam boiler is controlled by a(n) ___.
A. pressure control
B. air flow interlock
C. flame scanner
D. programmer

______ **15.** Water hammer in steam lines is caused by ___.
A. low steam pressure
B. high steam pressure
C. condensate in the line
D. a sudden drop in plant load

________________ **16.** A manometer measures ___.
A. flue gas temperature
B. volume of fuel oil flow
C. atmospheric pressure
D. difference in pressure between two points

________________ **17.** The best time to blow down a boiler is when it is ___.
A. at its peak load
B. at 75% of its peak load
C. at its lowest load
D. being taken out of service

________________ **18.** Boilers that are in dry layup have trays of ___ put in the steam and water drums to absorb moisture.
A. calcium chloride
B. hot soda lime
C. silica gel
D. potash

________________ **19.** To obtain complete combustion of a fuel, mixture, atomization, temperature, and ___ are required.
A. time
B. carbon dioxide
C. flue gas
D. hot refractory

________________ **20.** Oxygen in the boiler causes ___.
A. scale
B. pitting
C. foaming
D. carryover

________________ **21.** When a boiler low water alarm is ringing, the operator should ___.
A. call the chief engineer
B. increase the firing rate
C. decrease the feedwater
D. secure the fires and feedwater

________________ **22.** A furnace explosion can be prevented by ___.
A. checking the water level once a shift
B. testing the safety valves once a month
C. purging the furnace after ignition failure
D. testing the safety valve regularly

_______________ **23.** A lead sulfide cell, which is used for flame detection, is sensitive to ___.
A. infrared light
B. visible light
C. ultraviolet light
D. temperature increases

_______________ **24.** A pop-off safety valve is a ___ type.
A. hand and lever
B. pin and disc
C. deadweight
D. spring-loaded

_______________ **25.** To test a low water fuel cutoff using an evaporation test, the boiler operator must ___.
A. close the condensate return valve
B. secure all of the feedwater going to the boiler
C. completely drain the regulator chamber
D. have an inspector on hand

_______________ **26.** The ___ Hazard Signal System provides container label information about hazardous material.
A. ASME
B. NFPA
C. EPA
D. ANSI

_______________ **27.** To prevent a vacuum from forming in a boiler that is coming off-line, ___.
A. open the boiler vent
B. blow down the boiler
C. dump the boiler
D. pop the safety valve

_______________ **28.** The viscosity of fuel oil is the measurement of its ___.
A. Btu content
B. fire point
C. flash point
D. internal resistance to flow

_______________ **29.** Feedwater is treated chemically before it enters a boiler to ___.
A. prevent foaming
B. eliminate blowing down the boiler
C. change the scale-forming salts to a sludge that will raise the boiler water temperature
D. change scale-forming salts to a nonadhering sludge

_______________ **30.** Complete combustion is defined as burning all fuel using ___.
A. the theoretical amount of air
B. a minimal amount of excess air
C. no excess air
D. CO_2 and CO

_______________ **31.** When testing a safety valve by hand, there should be at least ___ psi of pressure in the boiler.
A. 1
B. 5
C. 15
D. 160

_______________ **32.** The water column on a low pressure steam boiler ___.
A. reduces fluctuation in water level to prevent carryover
B. reduces the turbulence of water in the gauge glass
C. provides a place to install the bottom blowdown valve
D. is never used

_______________ **33.** Safety valves on low pressure boilers ___.
A. can be tested only by hand
B. can be tested only by pressure
C. can be tested by hand or by pressure
D. should never be tested

_______________ **34.** Before laying up a boiler, the boiler operator must ___.
A. notify the boiler inspector
B. remove the boiler certificate
C. thoroughly clean the fire and water sides
D. only clean the water side because soot acts as an insulator

_______________ **35.** A scotch marine boiler is a ___ boiler.
A. watertube
B. cast iron
C. cast iron firetube
D. firetube

_______________ **36.** When performing a hydrostatic test on a boiler, ___.
A. all safety valves are removed or gagged
B. the bottom blowdown is opened
C. the main steam stop valve is opened
D. the vent is opened

____________________ **37.** ___ is the difference in pressure between two points of measurement that causes air or gases to flow.

A. Draft
B. Feedwater
C. Condensate
D. Carryover

____________________ **38.** No. 6 fuel oil has ___ viscosity compared to No. 2 fuel oil.

A. no
B. high
C. similar
D. low

____________________ **39.** ___ is heat transfer that occurs when molecules in a material are heated and the heat is passed from molecule to molecule through the material.

A. Combustion
B. Conduction
C. Condensation
D. radiation

Boiler Operator Licensing

Sample Exam 2

Name ______________________ **Date** ______________

True-False

T F **1.** To prevent water hammer, water should be removed from all steam lines.

T F **2.** Pressure on water does not change the boiling point of the water.

T F **3.** A Bourdon tube in a steam pressure gauge is filled with live steam when it is operating.

T F **4.** A steam trap is a device that removes air and condensate without loss of steam.

T F **5.** Treated water is used in a boiler to prevent pitting of the boiler metal.

T F **6.** When dumping a boiler, there should be at least 2 psi of steam in the boiler to force water from it.

T F **7.** The boiler vent should be open when filling the boiler with water.

T F **8.** A pressure control must be mounted in a vertical position to ensure accurate operation.

T F **9.** A burner should always start up in high fire to ensure enough fuel for ignition.

T F **10.** Carrying too high a water level in a boiler could lead to water hammer.

Multiple Choice

______________ **1.** On boilers that have both quick-opening and screw blowdown valves, the quick-opening valve must be located ___.

A. after the screw valve
B. at the NOWL
C. closest to the boiler and followed by the screw valve
D. on the boiler vent line

______________ **2.** Entering a boiler for maintenance may require a(n) ___.

A. OSHA certification
B. ASME license
C. boiler operator license
D. confined space permit

_______________ **3.** The Bourdon tube in a steam pressure gauge is protected by a ___.
A. steam trap
B. siphon
C. steam strainer
D. stopcock

_______________ **4.** Draft is measured with a ___.
A. pyrometer
B. hydrometer
C. manometer
D. thermocouple

_______________ **5.** A steam boiler is blown down to ___.
A. lower the oxygen level
B. test the safety valve
C. remove sludge and sediment
D. clean the feedwater lines

_______________ **6.** A compound gauge indicates ___.
A. differential pressure
B. pressure or vacuum
C. absolute pressure
D. the sum of the pressure on two boilers

_______________ **7.** Atomization of fuel oil in a rotary cup burner is caused by the ___.
A. rotating cup and secondary air
B. rotating cup and primary air
C. pressure of the fuel oil
D. secondary and primary air

_______________ **8.** A low pressure steam boiler has a maximum allowable working pressure (MAWP) of up to ___ psi.
A. 15
B. 35
C. 100
D. 160

_______________ **9.** In a firetube boiler, the heat and gases of combustion pass ___.
A. through the tubes
B. around the tubes
C. through the condenser
D. between the cast iron sections

_______________ **10.** Safety valve connections must be made according to the requirements of the ___.
A. fire department
B. shift foreman
C. American Society of Mechanical Engineers (ASME)
D. American Boiler Manufacturers Association (ABMA)

______________ **11.** Boiler fittings are necessary for safety and ___.
A. appearance
B. efficiency
C. cosmetic purposes
D. steam production

______________ **12.** The most important fitting on a boiler is the ___.
A. low water fuel cutoff
B. feedwater regulator
C. superheater
D. safety valve

______________ **13.** A safety valve ___.
A. controls the boiler operating range
B. controls high or low water
C. prevents the boiler from exceeding its MAWP
D. controls high and low fire

______________ **14.** Safety valves are designed to ___.
A. open slowly to prevent water hammer
B. pop open
C. open only by hand
D. open slowly and then close with no drop in pressure

______________ **15.** A steam pressure gauge is calibrated in ___.
A. pounds per cubic inch
B. inches of water column
C. pounds per temperature increase
D. pounds per square inch

______________ **16.** Bottom blowdown lines on a watertube boiler are located on the ___.
A. feedwater pump
B. bottom of the mud drum
C. fuel oil pump inlet
D. lowest part of the combustion chamber

______________ **17.** When blowing down a boiler equipped with a quick-opening valve and a screw valve, the quick-opening valve is ___.
A. opened first and closed last
B. opened last and closed first
C. opened first and closed first
D. opened as the operator prefers

______________ **18.** If a quick-opening valve is used as a bottom blowdown valve, it must be located ___.
A. between the boiler and the screw valve
B. furthest from the shell of the boiler
C. outside the check valve
D. on the feedwater line

_______________ **19.** Surface tension on the water in a steam and water drum is increased by ___.
A. impurities that float on the surface of the water
B. inexperienced boiler operators
C. very soft water
D. a high steam load

_______________ **20.** The surface blowdown line on a boiler is located at the ___.
A. MAWP
B. NOWL
C. lowest heating surface
D. highest heating surface

_______________ **21.** Draft is defined as a difference in pressure that causes ___.
A. air or gases to flow
B. a balance of pressure
C. a back pressure
D. feedwater to flow

_______________ **22.** Natural draft is produced by a(n) ___.
A. forced draft fan
B. induced draft fan (used in larger plants)
C. difference in temperature of a column of gas inside the stack from a column of air outside the stack
D. combination of fans

_______________ **23.** A draft fan located between a boiler and stack is used in a(n) ___ draft system.
A. natural
B. forced
C. induced
D. natural-forced

_______________ **24.** Duplex strainers are found on ___ of the fuel oil pump.
A. the suction side
B. the discharge side
C. both the suction and discharge sides
D. gas side

_______________ **25.** Fuel oil heaters must be used when burning No. ___ fuel oil.
A. 1
B. 2
C. 4
D. 6

______________ **26.** ___ oversees adherence to codes involving the construction and repairs of boilers and pressure vessels.
A. The EPA
B. The National Board
C. OSHA
D. The boiler inspector

______________ **27.** The ___ valve controls the firing rate in a high pressure gas system.
A. solenoid
B. manual reset
C. check
D. butterfly

______________ **28.** The gas pressure regulator in a low pressure gas system reduces the gas pressure to approximately ___ psi.
A. 0
B. 5
C. 10
D. 15

______________ **29.** Boilers are equipped with a combination burner for ___.
A. higher pressure operation
B. more flexible operation
C. safer operation
D. higher rates of combustion

______________ **30.** Draft is measured in ___.
A. pounds per square inch
B. inches or tenths of an inch of a vertical water column
C. inches or tenths of an inch of mercury
D. ounces per square inch

______________ **31.** ___ draft uses a fan before and after the boiler.
A. Forced
B. Induced
C. Natural
D. Combination forced and induced

______________ **32.** The amount of draft available in a natural draft system is dependent on the ___.
A. water pressure
B. height of the stack
C. fly ash in the gases of combustion
D. types of fans used

______________ **33.** Mechanical draft is produced by the ___.
A. height of the stack
B. diameter of the stack
C. power-driven fans
D. steam jets

________________ **34.** Mechanical draft can be classified as ___.
A. pressurized
B. natural
C. regenerative
D. forced or induced

________________ **35.** The induced draft fan is located ___.
A. between the boiler and the stack
B. at the base of the stack
C. at the burner
D. in the first pass of the gases of combustion

________________ **36.** The type of draft used in a fireplace is ___ draft.
A. mechanical
B. natural
C. forced
D. combination

________________ **37.** When the temperature of fuel oil is raised, its viscosity ___.
A. remains the same
B. is raised
C. is lowered
D. cannot be affected by heat

________________ **38.** The temperature at which fuel oil gives off a vapor that ignites readily when exposed to an open flame is its ___ point.
A. flash
B. fire
C. pour
D. viscosity

________________ **39.** The lowest temperature at which fuel oil will flow is its ___ point.
A. flash
B. fire
C. pour
D. viscosity

________________ **40.** A(n) ___ is burner control equipment that monitors the burner start-up sequence and the main flame during normal operation.
A. flue gas analyzer
B. modulation control system
C. flame safeguard system
D. air flow balancer

Boiler Operator Licensing

Sample Exam 3

Name ______________________________ **Date** ____________________

True-False

T F **1.** Blowdown of a boiler is not required if proper feedwater treatment is used.

T F **2.** A low water fuel cutoff shuts off the fuel to the burner when a low water condition exists.

T F **3.** A water column reduces water turbulence to allow a more accurate reading in the gauge glass.

T F **4.** Blowing down a water column and gauge glass too frequently can cause a false water level.

T F **5.** A stop valve on the feedwater line closest to the shell of a boiler allows repair of the check valve without dumping the boiler.

T F **6.** A vacuum pump and condensate tank are designed to discharge air and pump water.

T F **7.** Makeup water is added when boiler water is above the NOWL.

T F **8.** Most makeup water contains some scale-forming salts.

T F **9.** A globe valve should never be used as a main steam stop valve.

T F **10.** When open, gate valves offer no restriction to flow.

Multiple Choice

______________ **1.** The purpose of a ON/OFF pressure control is to ___.
A. open and close the water column
B. regulate air flow
C. regulate fuel flow
D. start and stop the burner on steam pressure demand

______________ **2.** A steam siphon ensures that ___ does not enter the boiler pressure control.
A. steam
B. water
C. air
D. gas

__________ **3.** A flame scanner is sensitive to ___.
A. heat
B. temperature
C. pressure
D. infrared rays

__________ **4.** In the event of a flame failure, the programmer ___.
A. sends more fuel to start a new firing cycle
B. secures the fuel and purges the furnace
C. vents the firetubes
D. stops the induced draft fan

__________ **5.** Pressure gauges are calibrated in pounds per ___.
A. square foot
B. cubic inch
C. vertical inch
D. square inch

__________ **6.** A boiler steam pressure gauge must be connected to the ___ of a boiler.
A. lowest part of the steam side
B. lowest part of the water side
C. highest part of the steam side
D. highest part of the water side

__________ **7.** Vacuum gauges are calibrated in inches of ___ atmospheric pressure.
A. mercury below
B. water below
C. mercury above
D. water above

__________ **8.** Boiler water must be treated to prevent ___.
A. formation of scale
B. overpressure
C. flame failure
D. feedwater pump failure

__________ **9.** Oxygen in a boiler causes ___.
A. scale
B. pitting of boiler metal
C. caustic embrittlement
D. carryover

__________ **10.** Carryover can lead to ___.
A. excess fuel oil temperature
B. water hammer
C. gas overpressure
D. air in the feedwater lines

________________ **11.** Oxygen present in water is commonly removed by ___.
A. adding lead sulfide
B. heating the feedwater
C. using the bottom blowdown valves
D. using the surface blowdown valves

________________ **12.** When taking over a shift, the boiler operator first checks the ___.
A. boiler room log
B. fuel oil supply
C. water level on all boilers that are on the line
D. bottom blowdown valves

________________ **13.** The purpose of a flame safeguard system is to protect the boiler from ___.
A. improper fuel oil pressure
B. a possible furnace explosion
C. exceeding its MAWP
D. starting in low fire

________________ **14.** After verifying the proper boiler water level, the burner may be started after ___.
A. filling the vacuum tank
B. venting the drum condenser
C. purging the furnace
D. purging the water tubes

________________ **15.** As a boiler is cooling down, the boiler operator must maintain the ___.
A. NOWL
B. excess air
C. normal feedwater temperature
D. 10% CO_2 reading

________________ **16.** Boilers that are out of service for an extended period of time ___.
A. are stored with an NOWL
B. are filled with oxygen
C. require more than 8 psi in the boiler
D. require proper layup procedures

________________ **17.** If there is a danger of a boiler freezing, the boiler should be laid up ___.
A. with a light fire
B. by using steam from the header to keep the boiler warm
C. dry with all water removed
D. using an antifreeze

________________ **18.** A low water fuel cutoff ___.
A. shuts the burner down
B. increases steam pressure to the load
C. adds water to the boiler
D. purges the boiler furnace

__________ **19.** A boiler that has had a low water condition should be ___.
A. thoroughly examined for signs of overheating
B. thoroughly examined for scale buildup
C. brought up to full steam pressure to test for leaks
D. brought up to the MAWP as soon as possible

__________ **20.** Flame scanners are ___.
A. equipped with a flame sensor
B. found on fuel oil strainers
C. found only on high pressure boilers
D. found only on low pressure boilers

__________ **21.** A furnace explosion can be caused by ___.
A. excess draft
B. a low steam pressure condition
C. an overheated furnace
D. an accumulation of fuel vapors

__________ **22.** Class ___ fires burn oil, grease, paint, and other flammable liquids.
A. A
B. B
C. C
D. D

__________ **23.** An unsafe condition should be reported to the ___.
A. shift operator
B. immediate supervisor
C. fire department
D. plant personnel department

__________ **24.** To start and sustain a fire, ___ are required.
A. fuel, heat, and CO_2
B. fuel, combustible matter, and CO_2
C. fuel, heat, and nitrogen
D. fuel, heat, and oxygen

__________ **25.** A fire caused by the ignition of wood, paper, or textiles is a Class ___ fire.
A. A
B. B
C. C
D. D

__________ **26.** A ___ is the use of locks, chains, or other physical restraints to prevent the operation of equipment.
A. lockout
B. tagout
C. permit authorization
D. restriction form

________________ **27.** A low water fuel cutoff should be blown down ___.
- A. daily or more often
- B. monthly
- C. every six months
- D. during annual inspections

________________ **28.** The ___ senses the water temperature in a hot water boiler.
- A. low water fuel cutoff
- B. aquastat
- C. vaporstat
- D. feedwater regulator

________________ **29.** A drop in water level in a steam boiler causes the automatic feedwater regulator to ___.
- A. close the fuel oil solenoid valve
- B. increase the flow of feedwater
- C. reduce the flow of feedwater
- D. increase the fuel oil to the burner

________________ **30.** A ___ is a regulation or minimum requirement.
- A. standard
- B. code
- C. technical bulletin
- D. recommendation

________________ **31.** In the safe operation of a steam boiler, the most important rule to follow is to ___.
- A. fill out the boiler room log
- B. perform bottom blowdowns regularly
- C. read the operation manual daily
- D. maintain the proper boiler water level at all times

________________ **32.** ___ valve is allowed between the boiler shell and the safety valve.
- A. An os&y
- B. A lever
- C. A check
- D. No

________________ **33.** A low pressure steam boiler safety valve setting cannot exceed ___ psi.
- A. 15
- B. 20
- C. 25
- D. 160

________________ **34.** ___ allow for the movement caused by expansion and contraction of steam lines from heating and cooling.
- A. Lagging joints
- B. Convection valves
- C. Bellow plates
- D. Expansion bends

__________ **35.** In a flame safeguard firing cycle, before the fuel oil valve can open, the flame safegaurd programmer must first ___.
A. purge the furnace
B. prove pilot ignition
C. prove water level
D. start feedwater flow

__________ **36.** The low water fuel cutoff should be tested ___.
A. with the burner firing
B. with the burner off
C. by removing the flame scanner
D. once a heating season

__________ **37.** If the low water alarm starts ringing and no water is visible in the gauge glass, ___.
A. secure the burner and feedwater
B. blow down the water column
C. add feedwater quickly and reduce the firing rate
D. start another feedwater pump

__________ **38.** The most accurate method used to determine the amount of feedwater treatment required is ___.
A. boiler water analysis
B. measurement of feedwater pressure
C. measurement of blowdown pressure
D. checking boiler water temperature

__________ **39.** The boiler vent should be kept open when ___.
A. the boiler is warmed up
B. the boiler has maximum line pressure
C. steam is blowing out of the vent
D. the boiler is cut off the line

__________ **40.** To perform an evaporation test on the low water fuel cutoff, ___.
A. have the chief engineer in attendance
B. close the main steam stop valve
C. secure all feedwater and makeup water going to the boiler
D. completely drain the low water fuel cutoff float chamber

Chapter 

Boiler Operator Licensing

Sample Exam 4

Name ______________________________ **Date** ______________________

True-False

T F **1.** Steam boilers that are out of service for an extended period require proper layup.

T F **2.** The dry method of boiler layup requires the boiler to be left open with air circulating to keep the boiler drums and tubes dry.

T F **3.** Chemical treatment of boiler water is not required for wet layup if the water has been deaerated.

T F **4.** The flash point of fuel oil is lower than its fire point.

T F **5.** The fire point of fuel oil is the minimum temperature at which fuel oil burns continuously.

T F **6.** No. 6 fuel oil burns with a clean flame when it is not heated.

T F **7.** If the gasket leaks on a duplex strainer, air could be drawn into the fuel oil lines.

T F **8.** Dirty strainers produce low suction readings.

T F **9.** Rotary cup burners can only burn No. 2 fuel oil.

T F **10.** Air used to atomize fuel oil is primary air.

T F **11.** Combination burners are used only in high pressure plants.

Multiple Choice

______________ **1.** ___ should be used to clean the inside of a gauge glass.
A. A wire brush
B. A cloth wrapped around a wooden dowel
C. A screwdriver wrapped with a paper towel
D. Sandpaper mounted on a steel rod

______________ **2.** A ___ is an accepted reference or practice.
A. standard
B. code
C. recommendation
D. law

_______________ **3.** An aquastat is an automatic device that senses ___.
A. water pressure
B. steam pressure
C. water temperature
D. steam temperature

_______________ **4.** At pressures higher than atmospheric pressure, water ___.
A. boils at less than 212°F
B. boils at exactly 212°F
C. boils at more than 212°F
D. does not boil

_______________ **5.** The number of blowdowns a boiler requires is determined by ___.
A. boiler water analysis
B. checking stack temperature
C. boiler manufacturer data
D. steam flow in the plant

_______________ **6.** Complete combustion is the combustion of fuel with ___.
A. the theoretical amount of air
B. the minimum amount of excess air
C. no excess air
D. no smoke or CO_2

_______________ **7.** Oxygen in a boiler causes ___.
A. priming
B. foaming
C. pitting
D. carryover

_______________ **8.** No. ___ fuel oil has the highest heating value in British thermal units per gallon.
A. 2
B. 4
C. 5
D. 6

_______________ **9.** The amount of ___ draft produced is affected by the outdoor temperature.
A. natural
B. forced
C. induced
D. balanced

_______________ **10.** To prepare a boiler for inspection, the boiler should be ___.
A. at MAWP
B. on-line
C. cool, open, and thoroughly clean
D. ready to dump when the inspector arrives

__________ **11.** Information on the boiler nameplate includes the ___.
A. hours of operation remaining
B. amount of steam produced over the boiler's lifetime
C. boiler owner's name
D. MAWP

__________ **12.** A safety valve is commonly tested during normal plant operation by the ___.
A. chief engineer
B. state inspector
C. operator on duty
D. plant manager

__________ **13.** In a straight-tube watertube boiler, heat and gases of combustion pass through the ___.
A. tubes
B. furnace
C. fuel oil strainer
D. fuel heater

__________ **14.** A feedwater pump can become steambound if ___.
A. steam pressure is too high
B. water pressure is too high
C. steam pressure is too low
D. feedwater temperature is too high

__________ **15.** When burning No. 6 fuel oil, a(n) ___ is needed.
A. fuel oil heater
B. feedwater heater
C. air preheater
D. condensate cooler

__________ **16.** The height of the stack determines the amount of ___ draft.
A. forced
B. induced
C. combination forced and induced
D. natural

__________ **17.** The water column on a low pressure boiler is used to ___.
A. measure the water level
B. indicate steam pressure
C. reduce turbulence in the gauge glass
D. blow down the boiler

__________ **18.** To protect against a furnace explosion, ___.
A. keep the water side clean
B. purge the boiler after all flame failures
C. check the damper
D. test safety valves once a week

______________ **19.** A water column should be blown down ___.
- A. once a day
- B. once a month
- C. once a shift
- D. when taking a boiler off-line

______________ **20.** The amount of fire in a burner is controlled by a(n) ___.
- A. vaporstat
- B. operating pressure relief
- C. aquastat
- D. modulating pressure control

______________ **21.** A clogged duplex fuel oil strainer located on the suction side of a pump results in ___.
- A. a high discharge pressure
- B. a high reading on the suction gauge
- C. a high fuel oil temperature
- D. increased fuel oil consumption

______________ **22.** A fan located between the breeching and stack is a(n) ___ fan.
- A. stack
- B. breeching
- C. forced draft
- D. induced draft

______________ **23.** ___ is when water is carried over from a boiler into the steam lines.
- A. Purging
- B. Chattering
- C. Throttling
- D. Priming

______________ **24.** Impurities on the surface of the water in a steam and water drum are removed by a(n) ___.
- A. continuous blowdown
- B. intermittent blowdown
- C. surface blowdown
- D. bottom blowdown

______________ **25.** A main steam stop valve is most commonly a(n) ___ valve.
- A. check
- B. globe
- C. modulating
- D. os&y gate

______________ **26.** A change in the condensate return temperature can be caused by a malfunctioning ___.
- A. feedwater pump
- B. fuel oil pump
- C. steam trap
- D. fuel oil strainer

_______________ **27.** A pressure control on a steam boiler is used to ___.
- A. control operating fuel oil temperature
- B. control makeup water pressure
- C. control high and low fire
- D. start and stop the breeching cleanout

_______________ **28.** Pressure at the discharge side of a forced draft fan is ___ atmospheric pressure.
- A. less than
- B. more than
- C. the same as
- D. balanced at

_______________ **29.** Before dumping a boiler, ___.
- A. allow the boiler to cool
- B. call the inspector
- C. open the blowdown valve with 5 to 10 psi on the boiler
- D. open all surface blowdown valves

_______________ **30.** Water weighs approximately ___ lb/gal.
- A. 1.0
- B. 8.3
- C. 62.4
- D. 212

_______________ **31.** Foaming of boiler water results when the water is contaminated with foreign material that causes ___.
- A. an increase in makeup water added
- B. more blowdowns
- C. an increase in fuel consumption
- D. an increase in surface tension

_______________ **32.** Blowing down is most effective when the steam output is ___.
- A. at a high rate
- B. at a low rate
- C. at zero
- D. early in the shift

_______________ **33.** A water column must be located ___.
- A. on the right side of the boiler
- B. on the left side of the boiler
- C. on the front of the boiler
- D. at the NOWL

_______________ **34.** A low water fuel cutoff should be tested with an evaporation test ___.
- A. daily
- B. weekly
- C. monthly
- D. yearly

_______________ **35.** A pressure control is protected from live steam by a ___.
A. stop valve
B. try cock
C. siphon
D. check valve

_______________ **36.** A boiler produces black smoke when there is ___.
A. a low atmospheric pressure
B. an improper mixture of air and fuel
C. excess secondary air
D. excess primary air

_______________ **37.** In order to properly test a low water fuel cutoff, ___.
A. the burner must be OFF
B. there must be no pressure on the boiler
C. the burner must be firing
D. the fuel must be shut OFF

_______________ **38.** The purpose of an expansion tank in a hot water heating system is to allow for the expansion of ___.
A. water
B. hot air
C. air and steam
D. gas

_______________ **39.** A flame scanner is located on the ___.
A. main steam line
B. fuel oil line
C. front of the furnace
D. bottom blowdown line

_______________ **40.** A pressure control controls the boiler operating range by ___.
A. regulating the fuel oil pressure
B. starting and stopping the burner
C. regulating the water temperature
D. causing the safety valve to relieve pressure

_______________ **41.** A heavy accumulation of soot on a boiler heating surface results in ___.
A. loss of boiler efficiency
B. increased heat transfer
C. loss of fire
D. safety valve popping

Low Pressure Boilers

Chapter 13

Boiler Operator Licensing

Sample Exam 5

Name ______________________________ Date ____________________

True-False

T F **1.** A low water condition is a common cause of boiler explosions.

T F **2.** A steam pressure gauge is calibrated in pounds per square foot.

T F **3.** The Bourdon tube of a steam pressure gauge must be protected from live steam.

T F **4.** A bottom blowdown line returns water to the boiler.

T F **5.** The function of a steam trap is to remove air and condensate without the loss of steam.

T F **6.** Purging a furnace before firing can prevent a furnace explosion.

T F **7.** Smoke is a sign of incomplete combustion.

T F **8.** A furnace must be purged after any flame failure.

T F **9.** An induced draft fan is located on the front of a boiler.

T F **10.** When using natural draft, there is a limit to the amount of fuel that can be burned.

T F **11.** A forced draft fan is located in the breeching.

T F **12.** Oxygen in a boiler water is used to remove nonadhering sludge.

Multiple Choice

______________ **1.** An ignition failure can be detected by the ___.
A. low water fuel cutoff
B. feedwater makeup system
C. high fire control
D. flame scanner

______________ **2.** An automatic feedwater regulator is used to ___.
A. ensure the proper water level in the boiler
B. shut off the burner in the event of low water
C. control the burner operating range
D. modulate gas pressure to the burner

_________________ **3.** On a steam boiler, testing the operation of the safety valve by hand with the boiler under pressure should be performed ___.
A. at the start of each heating season
B. every 30 days
C. annually
D. when maximum water pressure is achieved

_________________ **4.** In a natural-circulation hot water heating system, excess water is collected in the ___.
A. aquastat
B. blowdown tank
C. expansion tank
D. relief valve tank

_________________ **5.** If ignition fails during burner startup, the ___ protects the boiler.
A. low water fuel cutoff
B. flame scanner
C. vaporstat
D. aquastat

_________________ **6.** The ___ is the automatic control that protects a boiler from being fired with a low water condition.
A. aquastat
B. whistle valve
C. low water fuel cutoff
D. flame scanner

_________________ **7.** An indication of incomplete combustion is ___.
A. increased combustion efficiency
B. black smoke from the stack
C. an increase in water level
D. higher steam pressure generated

_________________ **8.** A(n) ___ pressure control controls the amount of steam produced by varying the firing rate.
A. safety
B. ON/OFF
C. modulating
D. aquastat

_________________ **9.** The ___ operates by sensing boiler water temperature.
A. flame scanner
B. vaporstat
C. aquastat
D. pressure control

_________________ **10.** The purpose of performing a try lever test on a safety valve is to ___.
A. test valve operation
B. set the operating range of the boiler
C. test popping pressure
D. determine the NOWL

__________ **11.** A feedwater regulator depends on the operation of a ___.
A. water pressure solenoid
B. steam pressure orifice
C. water temperature gauge
D. float-controlled valve

__________ **12.** To increase ___, the height of the stack must be increased.
A. forced draft
B. induced draft
C. natural draft
D. combustion gas

__________ **13.** In most plants, the water column and gauge glass should be blown down at least ___.
A. each 8-hour shift
B. monthly
C. quarterly
D. annually

__________ **14.** When blowing down a low water fuel cutoff, ___.
A. more fuel is burned
B. the burner shuts off
C. steam pressure increases
D. condensate temperature increases

__________ **15.** The purpose of an aquastat is to ___.
A. control the operating temperature range of a hot water heating boiler
B. feed water to the system
C. maintain a minimum water level in the boiler
D. provide for the expansion of water

__________ **16.** A pressure-reducing valve in a hot water heating system ___.
A. maintains a specified water pressure in the boiler
B. shuts the burner OFF when there is low water in the boiler
C. reduces makeup water pressure
D. regulates the flow of steam from the boiler

__________ **17.** A relief valve is rated in ___ per hour.
A. normal operating water level
B. British thermal units
C. pounds per square inch
D. gallons

__________ **18.** Makeup water is ___.
A. treated feedwater fed to a burner
B. water added to a boiler
C. condensate return water
D. condensed steam

________ **19.** The water column outlets of a steam boiler are connected to the ___.
A. mud drum
B. blowdown line
C. water and steam section of the boiler
D. water tubes of the boiler

________ **20.** A low pressure boiler commonly receives makeup water from the ___.
A. boiler feed pump
B. injector
C. automatic city water makeup feeder
D. return lines

________ **21.** A pressure control is protected from live steam by a(n) ___.
A. mercury switch
B. shutoff valve
C. siphon
D. inspector's test valve

________ **22.** A safety valve is located ___.
A. in the blowdown line
B. on the main steam header
C. on the highest part of the steam side of the boiler
D. on the lowest part of the steam side of the boiler

________ **23.** In most states, the only type of safety valve allowed on a steam boiler is the ___ type.
A. lever
B. direct-loaded pop
C. spring-loaded pop-off
D. vacuum relief

________ **24.** A stop valve and a check valve are usually placed ___.
A. in the steam line leaving a watertube boiler
B. in the blowdown line of a firetube boiler
C. as close to the boiler as possible on the feedwater line
D. in line with the pressure control

________ **25.** The steam gauge is calibrated in ___.
A. pounds of pressure
B. inches of pressure
C. pounds per square inch
D. pounds per square foot

________ **26.** No. 6 fuel oil ___.
A. is atomized as it leaves the fuel oil tank
B. is mixed with No. 2 fuel oil before it is atomized
C. must be heated before it can be burned
D. has a lower heating value than No. 2 fuel oil

_______________ **27.** A compound pressure gauge indicates ___.
A. a difference in pressure
B. the sum of two line pressures
C. either pressure or vacuum
D. three different pressures at the same time

_______________ **28.** The purpose of a feedwater regulator is to ___.
A. sound an alarm if the water level in the boiler is too high
B. maintain the proper water level in the boiler
C. shut off the burner if the water gets too low
D. keep the steam pressure within the safe limits

_______________ **29.** A low water fuel cutoff should be blown down ___.
A. every day
B. once a month
C. at the end of the heating season
D. once a week

_______________ **30.** When a low water condition occurs, the low water fuel cutoff ___.
A. increases the feedwater supply to the boiler
B. sounds an alarm to warn of low water
C. shuts off the fuel supply and feedwater to the burner
D. places the boiler in low fire

_______________ **31.** To obtain an accurate reading of the water level in a gauge glass, ___.
A. add makeup water
B. use the bottom blowdown valve
C. blow down the gauge glass
D. use the surface blowdown line

_______________ **32.** To change the operating pressure of a steam boiler, ___.
A. adjust the fuel supply
B. open the bottom blowdown valve wide and then close it
C. adjust the pressure control
D. look at the gauge glass

_______________ **33.** A lead sulfide cell is used in a(n) ___.
A. pressure control
B. flame scanner
C. low water fuel cutoff
D. evaporation test

_______________ **34.** When performing a hydrostatic test on a boiler, the safety valves must be ___.
A. kept free to pop at a set pressure
B. opened
C. plugged at the discharge end
D. gagged or removed, and the opening must be blank flanged

____________________ **35.** A feedwater pump is used to ___.
A. pump water into a boiler
B. pump out the boiler room sump pit
C. circulate water through the hot water heating system
D. pump fuel oil to the fuel oil burner

____________________ **36.** The ___ can discharge to a blowdown tank.
A. gauge glass
B. safety valve
C. water column
D. blowdown line

____________________ **37.** The heating surface of a boiler is ___.
A. only in the furnace
B. only found on water tube boilers
C. that part of the boiler where water is located
D. the part of the boiler where water is on one side and gases of combustion are on the other side

____________________ **38.** The range of a steam pressure gauge is ___.
A. 1½ to 2 times the MAWP
B. the MAWP
C. not more than 6% over the MAWP
D. gauge pressure plus atmospheric pressure

____________________ **39.** For every foot of vertical piping filled with water, there is a hydrostatic pressure of ___ psi.
A. 0.433
B. 0.251
C. 0.144
D. 0.0433

Boiler Operator Licensing

Sample Exam 6

Name ______________________________ Date ____________________

True-False

T F 1. In a low pressure gas burner, the gas regulator reduces gas pressure to 0 psi.

T F 2. A refrigerant absorbs heat in the evaporator of a compression refrigeration system.

T F 3. The range of a steam pressure gauge should be at least 3 times the MAWP of the boiler.

T F 4. The temperature and pressure of a refrigerant in a compression refrigeration system is lowered when it is compressed.

T F 5. An absorption refrigeration system can use ammonia as the refrigerant.

T F 6. Sensible heat is the amount of heat that changes the measurable temperature of a substance but not its state.

T F 7. The mechanical energy of a compressor can be changed to heat energy in a refrigeration system.

T F 8. An aquastat is used on a compression tank to control flow to a circulating pump.

T F 9. A scotch marine boiler is a type of firetube boiler.

T F 10. A pressure control regulates the firing of a burner based on condensate return flow.

T F 11. A feedwater check valve is commonly installed between a stop valve and a boiler.

T F 12. In colder climates, No. 6 fuel oil requires the use of tank heaters.

T F 13. Cooling systems are commonly rated in British thermal units of cooling per hour.

T F 14. Perfect combustion occurs when a boiler has no soot in the gases of combustion.

T F 15. Scale is caused by an accumulation of minerals present in boiler water.

Multiple Choice

______________ 1. Combustion efficiency in a burner is controlled by ___.

A. primary air
B. secondary air
C. forced combustion gas
D. tertiary air

_______________ **2.** In a compression system, liquid refrigerant under high pressure is allowed to drop in pressure by the ___.
A. absorber
B. absorbent
C. metering device
D. condenser

_______________ **3.** The capacity of a safety valve is measured by the amount of steam that can be discharged per ___.
A. shift
B. minute
C. hour
D. blowdown

_______________ **4.** Heat added to a substance that changes its state without a change in temperature is ___ heat.
A. super
B. sensible
C. latent
D. mechanical

_______________ **5.** In a refrigeration system, heat is ___ when a fluid changes from a gas to a liquid.
A. decreased
B. absorbed
C. released
D. compressed

_______________ **6.** An ___ is printed material used to relay chemical hazard information from the manufacturer to the employee.
A. OSHA regulation
B. SDS
C. EPA specification sheet
D. ASME notice

_______________ **7.** Excessive water in a boiler can lead to ___.
A. scale deposits
B. water hammer
C. flame failure
D. foaming

_______________ **8.** ___ is a government regulatory agency that was established to control and abate pollution.
A. DOT
B. OSHA
C. ANSI
D. The EPA

_______________ **9.** An absorption cooling system does not include a(n) ___.
A. condenser
B. compressor
C. evaporator
D. generator

______ **10.** In a lithium bromide and water cooling system, refrigerant is heated with a steam coil in the ___.
A. evaporator
B. compressor
C. generator
D. condenser

______ **11.** A Bourdon tube is used in a(n) ___.
A. compression tank
B. evaporator
C. compressor
D. pressure gauge

______ **12.** In the low pressure zone of a compression refrigeration system, ___ by the refrigerant.
A. heat is absorbed
B. steam pressure is increased
C. lithium bromide is produced
D. ammonia vapors are generated

______ **13.** Soot buildup on heating surfaces acts as an ___ to prevent the transfer of heat.
A. evaporator
B. insulator
C. ionizer
D. lubricant

______ **14.** Absorption cooling systems commonly use ___ as a refrigerant.
A. R-134a only
B. R-134a and water
C. ammonia only
D. ammonia and water

______ **15.** Flame scanners can use a(n) ___ to sense infrared rays of a pilot light and main burner.
A. lead sulfide cell
B. solenoid
C. purge sensor
D. ultrviolet sensor

______ **16.** In a compression refrigeration system, high pressure vapor is converted from a gas to a liquid in the ___.
A. evaporator
B. compressor
C. generator
D. condenser

______ **17.** Oxygen is removed from boiler water by ___ the water.
A. adding minerals to
B. chilling
C. heating
D. adding CO_2 to

________________ **18.** In a compression refrigeration system, the refrigerant absorbs heat in the ___.
A. diverter valve
B. compressor
C. generator
D. evaporator

________________ **19.** Steam that has released its heat turns to ___.
A. evaporated steam
B. sensible steam
C. condensate
D. latent vapors

________________ **20.** A low water fuel cutoff should be tested ___.
A. daily
B. monthly
C. quarterly
D. annually

________________ **21.** Some ___ containing chlorofluorocarbons (CFCs) cause damage to the Earth's ozone layer.
A. lithium bromides
B. nitrogen fuels
C. steam generators
D. refrigerants

________________ **22.** The burning of all of the fuel in a burner using a minimum amount of air is ___ combustion.
A. perfect
B. theoretical
C. complete
D. incomplete

________________ **23.** The temperature at which fuel oil must be heated to burn continuously when exposed to an open flame is the ___ point.
A. fire
B. flash
C. ignition
D. thermal

________________ **24.** Water used as a medium in indirect cooling systems must be kept above ___°F.
A. 0
B. 22
C. 32
D. 212

________________ **25.** Combustibles in a Class A fire are ___.
A. grease and gasoline
B. paper and wood
C. paints and solvents
D. electrical equipment

__________ **26.** A furnace must be ___ after every flame failure.
A. cleaned
B. blown down
C. purged
D. pressurized

__________ **27.** The heating value of a fuel oil is expressed in ___.
A. tons of heating
B. therms
C. flame units
D. British thermal units

__________ **28.** A low pressure hot water boiler has a MAWP of ___ psi.
A. 5
B. 15
C. 25
D. 160

__________ **29.** There are no tubes in a ___ boiler.
A. cast iron sectional
B. vertical wet-top
C. Scotch Marine
D. multiple-pass dry top

__________ **30.** Pressure in the high pressure side of a compression refrigeration system is produced by the ___.
A. expansion valve
B. condenser
C. compressor
D. evaporator

__________ **31.** In a refrigeration system, heat is ___ when a fluid changes from a liquid to a gas.
A. produced
B. absorbed
C. released
D. radiated

__________ **32.** The most common medium used to transport heat from an area to be cooled is ___.
A. water
B. ammonia
C. freon and water
D. steam

__________ **33.** Boiler fittings are manufactured in accordance with the ___ code.
A. boiler certification
B. ASME
C. NFPA
D. OSHA

_____ **34.** A ___ blowdown is performed to reduce foaming of boiler water.
A. bottom
B. surface
C. priming
D. huddling

_____ **35.** Hot water boilers operating with a 250°F water temperature and ___ psi water pressure or less are classified as low pressure boilers.
A. 15
B. 100
C. 160
D. 212

_____ **36.** ___ in boiler water causes corrosion, rusting, and pitting.
A. Sulfates
B. Sulfites
C. Nitrogen
D. Oxygen

_____ **37.** A(n) ___ control system for a boiler controls the amount of steam produced by changing the burner firing rate.
A. modulating
B. ON/OFF
C. proving
D. indirect

_____ **38.** Safety valves are repaired by ___.
A. the operator as required
B. an authorized manufacturer representative
C. the boiler inspector
D. the plant manager

Chapter 

Boiler Operator Licensing

Sample Exam 7

Name ______________________________ **Date** ____________________

Essay

1. What is the first thing a boiler operator must do when taking over a shift?

2. What are the different ways of getting water into a boiler?

3. What is the function of a pressure control?

4. How are safety valves tested?

5. How often should safety valves be tested?

6. What is the function of a low water fuel cutoff?

7. How is a low water fuel cutoff tested?

8. When should a safety valve pop on a low pressure steam boiler?

9. At what pressure should a safety valve pop on a low pressure steam boiler?

10. What is meant by "purging the furnace"?

11. What is the most important valve on a boiler?

12. How often should a steam boiler have a bottom blowdown?

13. How is a flame scanner tested?

14. What methods can be used to determine the water level in a boiler?

15. What is the main cause of smoke?

16. What are the different types of draft used in boilers?

17. What is the function of a check valve on a feedwater line?

18. What is the function of a stop valve on a feedwater line?

19. What actions would be required if a safety valve was popping and the steam pressure indicated 30 psi on the boiler?

20. What are perfect, complete, and incomplete combustion?

21. What could cause a furnace explosion?

22. What are the steps necessary in preparing for a boiler inspection?

23. How often should a gauge glass be blown down?

24. What is the difference between a forced-circulation hot water heating system and a natural-circulation hot water heating system?

25. What procedure is followed when performing a hydrostatic test on a boiler?

26. What does the acronym NOWL stand for?

27. What does the acronym MAWP stand for?

28. What is the cause of carryover?

29. What is the difference between natural and mechanical draft?

30. What causes a feedwater pump to become steambound?

31. What is the difference between a fast gauge and a slow gauge?

32. What is the cause of foaming in a boiler?

33. What is the difference between a gate valve and a globe valve?

34. What is the difference between a watertube boiler and a firetube boiler?

35. Where should a water column be located?

36. When is the best time to give a steam boiler a bottom blowdown?

37. What is the function of a boiler vent?

38. What is used to control high fire or low fire in a boiler?

39. How can a boiler operator tell when an os&y valve is in the open position?

40. What is the function of a vacuum pump?

41. How often should the automatic city water makeup feeder be blown down?

42. What are the most commonly used types of steam traps?

43. What are two methods of testing for the proper function of a steam trap?

44. How often should a fuel oil strainer be cleaned?

45. What are two commonly used types of fuel oil burners?

46. What are the advantages of using a combination burner?

47. What are primary, secondary, and excess air?

48. What is the function of a flame safeguard system?

49. What is hydrostatic pressure?

50. What are three types of heat transfer that occur in a boiler?

51. What is the function of a backflow preventer?

52. Why are expansion bends sometimes required on steam lines?

53. What is an evaporation test?

54. What are the benefits of a boiler room log?

55. What is the function of an aquastat?

56. What is a compressor and how does it function?

57. What special safety precautions must be followed when entering a confined space?

58. How is a boiler operator license obtained?

59. What is the proper procedure to use when trying to stop a leak from a gauge glass washer?

60. Describe how and when the main steam stop valve should be opened when routinely starting a boiler and bringing it on-line with another boiler.

61. What is the proper procedure to follow if a boiler shuts down, there is no water in the gauge glass, and the feedwater pump is running dry?

62. Describe where the water level should be when preparing to start a boiler that has been off-line for several days. Why should the water level be at that level? Describe what must be done about the water level.